Loving and Beyond

GOODYEAR EDUCATION SERIES
Theodore W. Hipple, Editor

CHANGE FOR CHILDREN
Sandra Nina Kaplan, Jo Ann Butom Kaplan, Sheila Kunishima Madsen, Bette K. Taylor

CRUCIAL ISSUES IN CONTEMPORARY EDUCATION
Theodore W. Hipple

CURRENT STRATEGIES FOR TEACHERS: A RESOURCE FOR PERSONALIZING INSTRUCTION
Robert L. Gilstrap and William R. Martin

EARLY CHILDHOOD EDUCATION
Marjorie Hipple

ELEMENTARY SCHOOL TEACHING: PROBLEMS AND METHODS
Margaret Kelly Giblin

FACILITATIVE TEACHING: THEORY AND PRACTICE
Robert Myrick and Joe Wittmer

THE FOUR FACES OF TEACHING
Dorothy I. Seaberg

THE FUTURE OF EDUCATION
Theodore W. Hipple

LOVING AND BEYOND: SCIENCE TEACHING FOR THE HUMANISTIC CLASSROOM
Joe Abruscato and Jack Hassard

MASTERING CLASSROOM COMMUNICATION
Dorothy Grant Hennings

THE OTHER SIDE OF THE REPORT CARD
Larry Chase

POPULAR MEDIA AND THE TEACHING OF ENGLISH
Thomas R. Giblin

RACE AND POLITICS IN SCHOOL/COMMUNITY ORGANIZATIONS
Allan C. Ornstein

REFORMING METROPOLITAN SCHOOLS
Allan Ornstein, Daniel Levine, Doxey Wilkerson

SCHOOL COUNSELING: PROBLEMS AND METHODS
Robert Myrick and Joe Wittmer

SECONDARY SCHOOL TEACHING: PROBLEMS AND METHODS
Theodore W. Hipple

SOLVING TEACHING PROBLEMS
Mildred Bluming and Myron Dembo

TEACHING, LOVING, AND SELF-DIRECTED LEARNING
David Thatcher

VALUE CLARIFICATION IN THE CLASSROOM: A PRIMER
Doyle Casteel and Robert Stahl

WILL THE REAL TEACHER PLEASE STAND UP?
Mary Greer and Bonnie Rubinstein

SOCIAL STUDIES AS CONTROVERSY
R. Jerrald Shive

A YOUNG CHILD EXPERIENCES
Sandra Nina Kaplan, Jo Ann Butom Kaplan, Sheila Kunishima Madsen, Bette Taylor Gould

Loving and Beyond
Science Teaching
for the Humanistic Classroom

Joe Abruscato
University of Vermont

Jack Hassard
Georgia State University

GOODYEAR PUBLISHING COMPANY, INC.
Santa Monica, California

Library of Congress Cataloging in Publication Data
Abruscato, Joe.
  Loving and beyond.

  (Goodyear education series)
  1. Science—Study and teaching. I. Hassard, Jack, joint author. II. Title.
Q181.A27   507   75-19568
ISBN 0-87620-535-X

Library of Congress Catalog Card Number: 75-19568

ISBN: 0-87620-535-X
Y-535X-2

Current Printing (last digit):
10 9 8 7 6 5 4 3 2

Sherri Butterfield   *Editor*
Linda Sanford Higgins   *Designer*
Kitty R. Anderson   *Illustrator*
Alice Harmon   *Production Artist*
Janice Gallagher   *Project Supervisor*
Pamela Morehouse   *Cover Illustrator*

Printed in the United States of America

Dear Fred Schlessinger:

This book is dedicated in your name to all the fine teachers around the country and around the world who are willing to dream of a better life for children and youth and are also willing to roll up their sleeves and do the hard work necessary to make a difference. The world has never been short of dreamers or doers, but it has been short of dreamer-doers.

Thank you for helping us to recognize our strong points, strengthen our weaknesses, and remember our humanness. Thank you, teacher and friend.

Contents

Teacher Stuff: Methods for Planning and Teaching in a Humanistic Classroom

Kid Stuff: Hugging a Tree and 80 Other Science Activities for Children

A light breeze kissing and teasing the meadow grass challenges the aerodynamic skill of a butterfly as it tries for a landing on a beautiful yellow flower. The gossamer-winged creature gently touches down for a brief visit . . . and then it is airborne once more carrying the almost magical yellow pollen that continues the amazing cycle played out millions of times a day as life begets life on our planet. The beautiful butterfly, the yellow flower, the swaying meadow grass, the furred and winged creatures, large and small, all joined together in a web of life, form but one dimension of the group of processes we call science. For science relates the plant world, the animal world, the mysteries of space, and the vagaries of climate with the processes of geology, chemistry, and physics. Science can be unifying, action-packed, mysterious, and wonderful for us all and especially for children.

The child is probably the best ''doer'' of science, for he need not scrape away the layers of prejudice and cynicism that form about each of us as we live yet another day. The child is curious, skeptical, and full of the natural drive and enthusiasm that we can only remember . . . and never regain. It is to that child that this book is directed. For he is us and we are he. His happiness is ours. And our future is his. To help him grow and to wonder and to become more than we can ever become is our hope.

Gentlepersons:

We each wish we could meet you personally, perhaps sit with you around a table sipping a steaming cup of hot dark chocolate topped with a great gob of whipped cream and simply talk. What a beautiful opportunity that would be, a chance to learn from you and a chance to share with you thoughts about children and science. We could talk about how kids learn or don't learn; we could talk about some ways to organize a classroom so that it becomes a rich environment in which children grow and learn; and we could talk about science activities that capitalize on the curiosity of the young. To meet each of you personally would be our pleasure, but it does seem a bit unlikely so we've put forth our part of the conversation within the pages of this book.

This book has been a great deal of fun for both of us. We've had a chance to express ourselves in some ways that are unique for a science methods book. So that you will know what you're getting into, you might first take a peek at the table of contents. Go ahead and look at it. We'll wait.

Did you notice that we've organized the first six chapters under the topic Teacher Stuff and the remaining chapters under Kid Stuff? We've focused on the basic methods of developing a humanistic learning environment for science in the first part. In the second part, we've provided science activities you'll be able to draw upon as a resource for your teaching.

At the end of each chapter in the book, we sit back and reflect a bit on the chapter. We call these reflections Conversation and we hope they will encourage you similarly to think about what we've presented.

If you get so interested in the ideas of any chapter from 1 to 6 that you wish to press ahead and do some additional outside reading, try the Reference Shelf at the end of those chapters. We thought you might want to challenge yourself on some of the things in the first six chapters so we developed a special page called Workshops at the end of each chapter. The Workshops page has some provocative questions and also one special activitiy called Center Stage. You might have fun doing Center Stage as a small group activity with other teachers or teachers in training.

At the end of this book is a rip-off page. We'd like you to use it to let us know how much you liked (or disliked) the book. If you have some science activity that you've tried successfully with children, let us know about it. Perhaps we can share it with others.

We wish you the best of luck, and we hope you enjoy working with this book.

Joe Abruscato
College of Education
University of Vermont
Burlington

Jack Hassard
School of Education
Georgia State University
Atlanta

Acknowledgments

Many people have contributed their help and inspiration to us while we developed this book. Ted Colton shared with us his ideas on how children learn science and provided us with many of the environmental science activities. We thank him also for many of the photographs that appear in the book. Charles Rathbone's and Frank Watson's ideas helped us shape the section on webbing. We give special thanks to Anita Bradford, Peggy Daniel, and Carolyn Hinsucker for sharing their beliefs and approaches to science teaching and conversing about it in Chapter 5. We wish to express appreciation to Eliot Galloway, headmaster of the Galloway School, and to Bill McGarrah, principal of the Lake Harbin Elementary School, for allowing us to visit their schools to photograph their teachers and children as they engaged in a humanistic approach to learning. We visited Anita's and Carolyn's classrooms and thank Sharon Kilpatrick and Jane Hannon for making their classrooms available to us also.

Many pre- and in-science teachers were our students during this project and contributed greatly to the ideas and activities that appear here. Their help was invaluable to both of us. We cannot list all your names, but we know who you are and we think a great deal about you. We wish to thank Joe Reynolds for introducing us to Goodyear, Janice Gallagher, Production Editor, for the personal attention we received, and Sherri Butterfield for editing our manuscript. Their continued support and excitement about this project from start to finish were great. The skill of our typists, Birute Conley and Mary Lou Wasko, who translated our illegible writing into an accurate manuscript, has astounded us—thanks for your patience.

Teacher Stuff:

Methods for Planning
and
Teaching in a
Humanistic Classroom

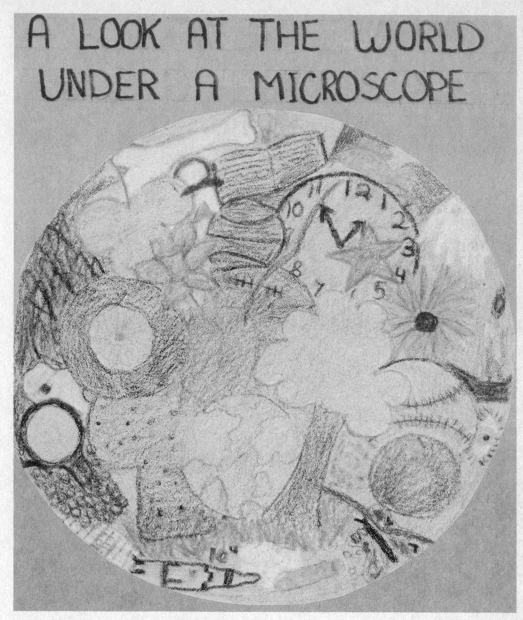

A LOOK AT THE WORLD UNDER A MICROSCOPE

Crayon.

1

Planning for a Humanistic Science Classroom

After Loving Comes...

Loving children is not enough. Most of us (you, me, and the teacher down the hall) would probably say that we love children, yet the suffocation of the young continues. Love is definitely not enough.

As teachers we must come to realize that our love for children must be channeled into actions. These actions are the hard work of teaching. They include planning, encouraging student progress, and finding ways of improving what we do. In this chapter we will be concerned with the planning phase of good science teaching, science teaching that provides children with experiences that help them to grow and to learn.

One explanation for the somewhat negative feeling adults may have about science is that many have learned to avoid science experiences throughout their schooling. Even that portion of our school population that goes on to further study tends to avoid energetically all contact with science. How can this tremendous avoidance phenomenon be explained? We think it is the result of certain things students learn in the schools with respect to instruction in science. They somehow learn that science is difficult, complex, and boring. Actually, science is far easier than reading, far less complex than learning the rules of English grammar, and potentially the most exciting part of the school day.

Science Is...

We were sitting in front of a living room fireplace, my daughter and I, savoring a moment that found each of us free from anxiety and able to reflect in any way we chose. Jennifer asked, "Daddy, why does the fire make that noise?"

I didn't have the faintest idea. I said, "Let's see if you and I can make some guesses about why the fire makes that noise. What do you think, Jenni?"

The next day we were on the shores of a lake of volcanic origin. Floating with the current were all sizes of grayish, foam-like things. Jenni risked touching one. It was not mushy. Conversely, she found it had real substance. These were, much to her astonishment, floating stones. Where did they come from? Why do they float? What are they called? We had an amazing discovery of pumice stone and its qualities. Then we discovered how easily it could be carved. Jenni collected a sackful to be saved for carving.[2]

Jenni was curious about the world around her. Her curiosity and subsequent discussion and exploration with her father beautifully capture the spirit of science. Science is, at its core, a *human* activity. It is a human enterprise through which *people* apply their curiosity by using various processes to acquire a fuller understanding of the natural world.

Children are born scientists who energetically test every facet of their environment (including their parents and teachers). This natural probing and questioning can be cultivated to its fullest possible limits as children engage in the study of science. The methods and activities described in this book should provide you with the basic ingredients for a successful science program, but much depends on you.

Science Teacher, Who Are You?

Are you alive? Are you in touch with the world around you? When you see a beautiful sunset, do special feelings enter your heart and mind? Have you ever walked through a pine forest on a warm day and smelled the aroma of pine? Have you ever petted a rabbit and felt its soft fur? Have you ever found a special flat-sided stone that you could get to skip across the top of a stream? Have you seen a newborn puppy or looked carefully at a single snowflake? Are you alive?

Do you ever wonder? Have you ever looked out at the universe on a crisp, star-filled night and wondered: Is anyone out there looking back at you? Have you ever watched a squirrel with hope in its heart make death-defying leaps from tree to tree and wondered: How does it do it without falling? Have you ever seen a porcupine and wondered: How in the world do porcupines mate? Answer: Very, very carefully.

Do you love kids? It's easy to like kids when they are good—cute boys and girls, smelling nice, saying "please" and "thank you," and "You're a great teacher, Mr. Smith or Mrs. Jones." But *love* is liking kids when they are bad, smell funny, say "gimme" instead of "please," or imply that:

1. Your mother is in some specific way related to the canine family.
2. You are the result of a relationship between your mother and father that was never formalized by reverent clergy or a justice of the peace.

If you are confused by 1 and 2 above, find a worldly friend who can translate them.

Liking Kids Is Easy,
But Are You Sure You Love Them?

Are you a winner? Do you believe in yourself? Do you believe in humanity? Do you believe that, in spite of wars and other bits of savagery from time to time, somehow, some way we can overcome? And about yourself, do you believe that you can climb the mountain, swim the river, slay the dragon, build a cathedral, . . .

. . . teach a child?

The great teacher is a winner when it comes to a belief in humanity. Shannon is right on target when he says:

The "if only" and "but" comments of the losers in teaching sound like this:

If only I had more textbooks.
If only I had smaller classes.
If only I had a free period.
If only they'd fire Miss Zing.
If only I had an overhead projector.
If only I had gifted students, more chalk, fewer meetings, no reports, a
    marriage—or a divorce—or a love affair.
If only we had a year to talk this over.

But they don't want to learn.
But they can't read.
But the principal won't let me.
But the superintendent is a bastard.
But parents don't care.
But I don't have time.
But that's too theoretical.
But sex is . . .
But it hasn't been researched.

Winners say:
Let's go!
Why don't we try this!
I think we can do it!
Come on in and see what we're doing!
The kids had a great idea!
I had a terrific weekend!
Maybe, this will work!
Don't worry about the principal, we can bring him around.

That's beautiful![3]

IF you are
  alive
    a wonderer
      a lover
        a winner,
then you are a person who can and will provide experiences for children through the medium of science that will help them to become all they are capable of becoming—socially, emotionally, and intellectually. Their day is tomorrow, and what they become will be a result of what happens to them today. The challenge is large, but alive, wondering, loving, and winning teachers will help their tomorrow be a little better than their today.

Humanism and Science: This We Believe

Some people resist the notion that the experience of studying science is a humanistic endeavor. We feel that it can be and should be for many reasons, including the following:

1. Science is a *human experience.* It involves humans looking out at their world.
2. Science usually involves a *cooperative human effort.* The scientist, alone, high in the ivory tower, is an inaccurate view of the scientific role.
3. The basic processes of science, such as discovering, valuing, and exploring, are applicable to *many of the human social problems* people face, problems that include achievement of social change and the improvement of interpersonal relationships.
4. Certain products of science as transmitted through technology can be used to *alleviate human suffering* resulting from poverty, disease, and illiteracy.
5. The essence of humanism, as we see it, is that each human being should be encouraged *to utilize his or her full human potential,* as well as intellectual and social potential. Science as a human endeavor provides opportunities for this to occur. Science provides us with valuable capabilities for investigation and a responsibility to use these capabilities for the benefit of others.

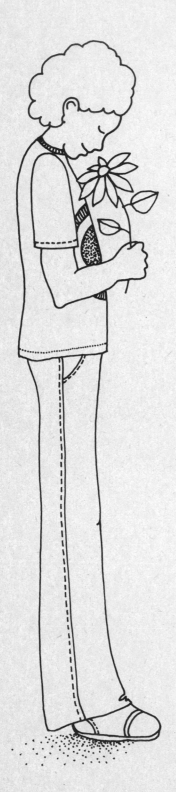

Let's consider in a little more detail the relationship of developing human potential and science. As we will note elsewhere in this book, the child is the best doer of science. The child's natural curiosity about all that is outside him or her and all that is within captures the spirit of science. The no-holds-barred questioning: How? Why? When? Where? What if . . .? The school environment can foster this type of questioning or it can stifle it.

Bringing these questions to bear on our environment and on ourselves is consistent with the spirit of science and additionally teaches us to raise questions about everything. This questioning attitude liberates us from our complacency. The humanist is seldom content with the status quo because things can be better. Science provides us with a way of looking beyond the way things are and leads us to question.

A moving story that has captured the imagination of millions of people around the world in recent years is Richard Bach's *Jonathan Livingston Seagull.* Through Jonathan we can see ourselves and begin to appreciate the relationship between the questioning spirit of science and the fulfillment of our human potential. Perhaps these excerpts will help you think about science and humanism.

Most gulls don't bother to learn more than the simplest facts of flight—how to get from shore to food and back again. For most gulls, it is not flying that matters, but eating. For this gull, though, it was not eating that mattered, but flight. More than anything else, Jonathan Livingston Seagull loved to fly.

This kind of thinking is not the way to make one's self popular with other birds. Even his parents were dismayed as Jonathan spent whole days alone, making hundreds of low-level glides, experimenting.

He didn't know why, for instance, but when he flew at altitudes less than half his wingspan above the water, he could stay in the air longer, with less effort. His glides ended not with the usual feet-down splash into

the sea, but with a long flat wake as he touched the surface with his feet tightly streamlined against his body. When he began sliding in to feet-up landings on the beach, then pacing the length of his slide in the sand, his parents were very much dismayed indeed.

"Why, Jon, why?" his mother asked. "Why is it so hard to be like the rest of the flock, Jon? Why can't you leave low flying to the pelicans, the albatross? Why don't you eat? Son, you're bone and feathers!"

"I don't mind being bone and feathers, mom. I just want to know what I can do in the air and what I can't, that's all. I just want to know."[4]

Jonathan put his curiosity into action, and went to work learning about himself and his world. Sometimes he failed.

By the time he passed four thousand feet he had reached terminal velocity, the wind was a solid beating wall of sound against which he could move no faster. He was flying now straight down, at two hundred fourteen miles per hour. He swallowed, knowing that if his wings unfolded at that speed he'd be blown into a million tiny shreds of seagull. But the speed was power, and the speed was joy, and the speed was pure beauty.

He began his pullout at a thousand feet, wingtips thudding and blurring in that gigantic wind, the boat and the crowd of gulls tilting and growing meteor-fast, directly in his path.

He couldn't stop; he didn't know yet even how to turn at that speed. Collision would be instant death.

And so he shut his eyes.

It happened that morning, then, just after sunrise, that Jonathan Livingston Seagull fired directly through the center of Breakfast Flock, ticking off two hundred twelve miles per hour, eyes closed, in a great roaring shriek of wind and feathers. The Gull of Fortune smiled upon him this once, and no one was killed.[5]

But he also discovered things about himself and the natural world. And each discovery helped him in reaching his full potential.

A single wingtip feather, he found, moved a fraction of an inch, gives a smooth sweeping curve at tremendous speed. Before he learned this, however, he found that moving more than one feather at that speed will spin you like a rifle ball . . . and Jonathan had flown the first aerobatics of any seagull on earth.

He spared no time that day for talk with other gulls, but flew on past sunset. He discovered the loop, the slow roll, the point roll, the inverted spin, the gull bunt, the pinwheel.[6]

And with this effort Jonathan began to move closer and closer to his goal. And his goal can be ours: becoming all we are capable of becoming.

When Jonathan Seagull joined the Flock on the beach, it was full night. He was dizzy and terribly tired. Yet in delight he flew a loop to landing, with a snap roll just before touch-down. When they hear of it, he thought, of the Breakthrough, they'll be wild with joy. How much more there is now to living! Instead of our drab slogging forth and back to the fishing boats, there's a reason to life! We can lift ourselves out of ignorance, we can find ourselves as creatures of excellence and intelligence and skill. We can be free! We can learn to fly![7]

Teacher and children can see in Jonathan a seagull—a person—who is not content with the status quo. A person who is constantly discovering new talents, new resources, and new ways of looking at ourselves and our capacities. Such is the spirit of science, of humanism, and of one lonely seagull who questioned.

The Traditional Science Curriculum

The science curriculum consists of all planned and unplanned science experiences that a child has during the year. The traditional science curriculum for children has been short on enjoyable experimentation and long on bookish pursuits. As a matter of fact, in many schools teachers and children have unfortunately become hostages to a textbook series.

In some schools the science textbook series has become the science curriculum, which is really a sad development. Although each of the major publishers has a series of well-designed, beautifully illustrated, and activity-laden science books, the utilization of the books as the total science program reflects poorly on our professional competence. If we find ourselves teaching science in a school that has adopted a textbook series, we should view the books as useful resources that are but one portion of the curriculum. The challenge to us is that of using them in a productive, positive way and not as a source of constraint on our imagination as weil as the children's.

Your Science Curriculum

The Philosophy

As you think about a year of science at the grade level you are most interested in teaching, a very fundamental question emerges: *To what extent should I specify the experiences (curriculum) in science that children should have?* At its more general level, all teachers at all grade levels face this question daily: it is the eternal question of student freedom versus teacher decision making. We have also struggled with this question as public school teachers and now as teachers of college students and teachers. We wish we could provide you with a definite answer, but we cannot. Perhaps we can help you think through this question by having you meet two teachers (see Figure 1).

As we all strive to provide more humane teaching-learning situations for children, we must guard against giving up our moral responsibility for the growth of children. None of us wants to be a "Closed Clara." We want children to develop their original thinking capabilities; we want to foster creative development; we want them to discover, value, and explore; but can they reasonably be expected to grow spontaneously in these ways in a classroom led by Open Oscar?

In our view of education, the science teacher has a responsibility to children and should provide neither the laissez-faire education of Open Oscar nor the authoritarian education of Closed Clara.

We hope you'll find a comfortable teaching philosophy somewhere between Oscar's and Clara's. In this book we'll be trying to help you become so competent in the methods and activities of science that you'll be a great teacher, regardless of the teaching style you develop. (But please, no more Claras and Oscars, thank you.)

Figure 1. Where are you in relation to Open Oscar and Closed Clara? Place a check mark along the line to indicate your position. Compare with others.

The Planning Process: Some Tools

Teachers sometimes avoid teaching science because of a fear that students will ask questions they are not ready for; a fear that experiments are too complicated to try; and a fear that science will use precious time that could be better utilized in trying to drill phonics or addition facts into the memories of their charges. Actually, the opposite is true, because a well-conceived science curriculum can engage students in so many action-oriented activities that they just may acquire enough enthusiasm about school to resist saying "I hate school" on their return home after a long day. Indeed the capitalization of science on the curiosity of the young can really help develop positive feelings toward teachers and schooling. Phonics and math facts are much easier to learn when the child genuinely likes being in school. Here then are some tools you can use as a curriculum planner.

Interest Lists

One approach to planning for science that you can try is the development of interest lists. To construct interest lists, you must begin to answer the question,

"What types of science experiences develop the highest amount of interest for the children that I teach?" The answer to the question need not be particularly complicated: a simple list will do. For an example, you might list the following:

1. Collecting things
2. Taking nature walks
3. Keeping animals
4. Growing plants
5. Doing simple electricity experiments
6. Studying the weather
7. Finding out about chemical reactions
8. Studying outer space

The list that you prepare will depend on such factors as the age and interests of the children and will be unique to you and your situation. Table 1 shows

Table 1.

Responses of 35 Teachers (5 at each Grade Level) to the Question:
What five science topics most interest the children at your grade level?

GRADE						
1	2	3	4	5	6	7
Air and wind	Animals—	Birds	Reproduction	Drugs	Animals	Animal
Magnets	living things	Dinosaurs	Chemicals	Dinosaurs	Conservation	behavior
Plant growth	Seasons	Space	Environmental	Human body	and ecology	Natural
Water	Human body	Insects	science	Fossils	Human body	phenomena
Colors	Outer space	Machines	Astronomy		Motors	Environmental
Living things	Ecology		Heat systems		Plants	issues
						Use of equip-
						ment
						Sex education
Animals	Mammals	Solids, liquids,	Space	Animals	Living	Human body
Magnets	Fish	gases	Weather	Outer space	organisms	Animals
Electricity	Plants	Soil	Electricity	Ecology	Space	Human and plant
Growing plants	Experiments	Fire pre-	Foods	Plants	Astronomy	reproduction
Collecting and	with water	vention	Water	Experiments	Energy	Ecology
observing		Teeth			Measurement	
insects		Five senses			Identification	
					of chemicals	
Animals	Magnets	Animal life	Ecological	Animals	Astronomy	Body system
Color mixing	Electricity	cycles	balance	Plants	Biology	Dissection
Magnifying	Dinosaurs	Plants	Diseases	Space	Organic	Microscope
glasses	Seeds, plants	Magnets and	Rocks and	Life in the	reactions	study
Measuring	Space	electricity	minerals	ocean	Ecology	Current events
Change of		Endangered	Electricity	Human body	Chemistry	in advancing
state		species				technology
		Pollution and				Nature study
		ecology				
Space, stars,	How, why	Space rockets	Animals	Matter	Animal life	Biology
planets	things work	Magnets	Plants	Weather	Plant life	Planets
Living things	Plants	Insects	Space	Astronomy	Metric system	Chemistry
Reproduction	Animals	Underwater	Rocks,	Animals	Solar system	Ecology
Rocks, minerals		plants	minerals	Magnetism	Conservation	Science fiction
Magnets			Electricity			
Electricity						
Animals	Animals	Rockets	Animals	Plants	Snakes	Cells
Plants	Space	Plants	Plants	Environment	Rockets	Animal and plant
Magnets	Prehistoric	Magnets	Space	Universe	Reproduction	reproduction
Rockets	animals	Insects	Astronomy	Insects	Insects	Geology
Rocks, minerals	Ocean life	Planets	Electricity	Rocks,	Transporta-	Astronomy
	Plants		Pollution	minerals	tion	

a collection of "interest lists" developed by five different teachers at each grade level from the first through the seventh. The best way of preparing your own list is to interview children at the age and grade level you are teaching or intending to teach. One measure of the extent of openness you will foster in your classroom is the degree to which you include the desires and interests of children in your planning.

There is much more said about science than is done. The pressure you feel as a teacher to teach the basic skills of reading and arithmetic may even make you feel guilty if you spend very much time thinking about science, let alone actually doing it. This is especially sad because the doing of action-oriented discovery science by children can be a great deal of fun for both you and your children. You may be a little "shy" about doing science if you've learned to fear it as a student. But don't feel bad: you are not alone. The important thing to keep in mind is that science for children does not need to be something for which you know all the answers. The essence of science is curiosity about the world around you. The answers to the questions that will be raised as a result of this curiosity are not to be found in books. The answers lie in the natural environment that you and the children will be exploring. And as you explore, you will discover one eternal truth: *The answers to the questions are questions!*

In teaching science our goal is not to teach children a multitude of facts or concepts but rather to help them become curious, to ask and to experiment. How then can you as a teacher assist students in becoming curious about their world? The answer depends upon your ability to pique their curiosity by providing opportunities for them to become actively engaged in the process of science. The processes of science can easily become the central focus for science instruction if you can create "learning opportunities" through *science activities.* There are two basic sources of science activities:

1. Student and teacher ideas
2. The ideas of others

Student and Teacher Ideas

Students ideas for activities they may wish to try usually come as a natural outgrowth of their day-to-day experiences. Your role is one of providing a classroom environment that will lead students to raise questions. A classroom that includes an aquarium, a terrarium, a collection of growing plants, and a few small animals will generate plenty of questions, such as:

1. "Why are the leaves turning yellow?"
2. "How do the fish breathe?" (Note: Fish really don't breathe.)
3. "Do rabbits lay eggs?"
4. "Where does dirt come from?"
5. "What's that green stuff floating on the water?"
6. "Does the hamster sleep at night?"

For some children *you* may have to be the one to raise some questions. You might wonder aloud about such things as:

1. What would happen if we planted some seeds upside down?
2. I wonder if fish could find their food if we covered the aquarium with a dark cloth?
3. I wonder how we could find out how much the gerbils weigh?
4. Do the gerbils spend more time sleeping than the hamsters?

The range of activities that you can engage in with children really depends on their imagination and your imagination. When you wish to develop science activities, just stop for a minute and look around and *wonder.*

The Ideas of Others

One excellent source of science activities are the science textbooks referred to above. Although their use as a total science curriculum is questionable, all of them include science activities. Simply acquire your own collection of science texts for elementary education and start searching out those activities that might be interesting to your children. Organize the activities you think might prove useful according to the goals or objectives that were discussed earlier.

Another source of ideas for science activities are articles written by and for teachers in such magazines and journals as:

Grade Teacher
CCM: Professional Magazines Inc.
22 West Putman Avenue
Greenwich, Connecticut 06830

Instructor
Instructor Publication, Inc.
Instructor Park
Dansville, New York 14437

Learning
Education Today Company
1255 Portland Place
Boulder, Colorado 80302

School Science and Mathematics
School Science and Mathematics Inc.
Box 246
Bloomington, Indiana 47401

Science and Children
National Science Teachers Association
1742 Connecticut Avenue, N.W.
Washington, D.C. 20009

The Science Teacher
National Science Teachers Association
1742 Connecticut Avenue, N.W.
Washington, D.C. 20009

Most or all of these periodicals should be available in your school's professional library (if one exists) or in the library of any local college or university. You may wish to subscribe to them yourself, thereby ensuring a continuing fresh supply of ideas.

In the Kid Stuff portion of this book we will be providing you with an organization of science activities that goes beyond a mere collection of things to do. Rather, the activities focus on three processes fundamental to learning science: discovering, valuing, and exploring. These three processes form the basis for turning your classroom into an exciting, action-oriented, loving, caring —in short, humanistic—classroom.

We hope you find the activities relevant and encourage you to acquire a much larger collection of activities through other sources.

And Now a Few Words about Objectives

In recent years an increasingly large amount of attention has been paid to the idea of trying to state the outcomes of the learning process and using these statements of outcomes (objectives) as the first step in instructional planning. Advocates of the position that the outcomes of learning should be prespecified point out that, in many instances, an overly open curriculum becomes no curriculum at all.

Teachers of science who are truly interested in the overall development of children can make some use of objectives as a planning tool if care is taken to keep the following things in mind:
1. Objectives-based activities should be only one component of the learning environment.
2. Children and teachers should share in the development of objectives.

By way of definition, we should point out that an objective that specifies an observable pupil behavior, or an observable product of pupil behavior, is often referred to as a "behavioral objective" or a "performance objective." A more general statement of an outcome of instruction is usually called a "goal" (aim). Here is an example of each.

Goal: Each child will learn about seeds, germination, and plant care.

Objective: Given a selection of seeds and potting soil, each child will plant a seed, keep the soil moist, and provide sunlight and water daily (or as needed) for newly sprouted plants.

We view behavioral objectives as just one tool for providing a successful learning environment. The challenge for us as teachers is to use them only in ways that will assist us in providing for the needs of children.

The "New" Curricular Materials

The need for an improvement in the manner through which children learn science is not a new development. In the past ten to fifteen years there has been a flurry of excitement as various groups of educators, private organizations, and agencies of the federal government have invested considerable time and resources in an attempt to improve the quality of science education. In the process various groups of disciples have emerged who have a special interest in "new" curriculum x, y, or z. You can hear that "curriculum x" is process oriented, "curriculum y" is more humanistic than x, or that "curriculum z" is process oriented, humanistic, and in addition, is divinely inspired.

This abundance of new curricular materials, although well intentioned and in most instances well developed, has been a source of considerable confusion to teachers and curriculum coordinators. Much time and effort is spent at the local school district level as teachers, curriculum coordinators, and even parental groups try to decide on curriculum x, y, or z. We view this process with some skepticism since the underlying assumption is that any single curriculum package will be best for any specific group of children. As we have said above, many of the "newer curricula" are excellent (as a matter of fact, we take some pride in noting that we've been personally involved in the development and implementation of a goodly number of them); however, the notion that one curriculum approach is the solution runs counter to everything we know and feel about the needs of children and teachers.

How then can we make use of these new curricula without adopting one for all the children in a class or a school. Well, you must be ready to take that first big step beyond loving: you've got to do some work. You will need to study the "newer curricula," and we really do mean study, not simply browsing over some brief outlines and nifty little activities. You will need to roll up your sleeves, write to the various agencies or curriculum publishers, get sample materials, actually try some of the activities, and begin selecting those specific parts of the available curricula that will work best for your children and you.

To save you some time, we've gathered a listing of 15 curriculum projects, brief descriptions, and addresses that you can use as a reference. They are shown in Table 2.

Teaching to us is a very creative act; and just as the creative artist sums up his impressions of reality in a beautiful work of art, we as teachers must take the time to gather together our impressions of the available curricular materials so that we can provide beautiful learning experiences for children.

Table 2.

Curriculum Projects You May Wish to Explore.

Name of Project	Address	Description
AAAS Science Supplement Environmental Education Enrichment Activities	M. T. McLean Environmental Education Austin Independent School District Austin, Texas 78752	This project was developed for children ages 6 through 11 and is interdisciplinary in nature. The emphasis of the project is in the following areas: 1. awareness of natural materials and processes in the school neighborhood; 2. observations of interactions and changes in the neighborhood environment; 3. consideration of time as a factor in natural process; 4. appreciation of the dependence of people on natural resources and processes.
Conceptually Oriented Program in Elementary Science (COPES)	New York University 4 Washington Place Room 261 New York, N.Y. 10003	An understanding of the nature of matter, both animate and inanimate, as a way of developing scientific literacy is the basis for this project. The materials developed for children ages 5 through 13 emphasize the exploration of the natural world using the basic underlying concepts of science.
Elementary Science Study (ESS)	Education Development Center 55 Chapel Street Newton, Mass. 02160	The purpose of this project is to enrich students' understanding about the world and themselves through the utilization of activities that require readily available materials. Over 55 science units have been developed for use by children in grades K through 12.
Essentia: Environmental Studies (ES)	The Evergreen State College Olympia, Wash. 98506	The materials developed through this project are designed as an interdisciplinary approach to environmental sciences for students in grades K through 12. The project materials are designed so that the teacher can use them independently or as a supplement to other curricular materials.
Individualized Science (IS)	Learning Research and Development Center 3939 O'Hara Street Pittsburg, Pa. 15260	An individualized science program for children in grades K through 8 has been the goal of this project. A management system has also been designed so that the child can help plan activities, and assess his or her learnings. The IS materials deal with a wide variety of concepts and processes in the many different areas of science explorations.
Intermediate Science Curriculum Study (ISCS)	ISCS Project 415 N. Monroe Street Room 705 Florida State University Tallahassee, Fla. 32301	This project has been designed for children from 12 through 15 years of age. The level 1 materials would be appropriate for students in the upper grades of middle school (6 or 7). The project materials allow the students to progress through various instructional pathways. The ISCS materials include independent study, laboratory investigations, discussion sessions, and self-pacing materials.
Man: A Course of Study (MACOS)	Social Studies Program Education Development Program 15 Mifflin Place Cambridge, Mass. 22138	This is an interdisciplinary social studies curriculum that incorporates a substantial number of activities that are "scientific" in nature. The elementary and middle school student for whom this curriculum was developed work through various activities as they explore three basic questions about human beings: "What is human about human beings?" "How did they get that way?" "How can they be made more so?"
Model Educational Program in Ecology	Laurel Ecology Center 1044 No. Hayworth Ave. Los Angeles, Calif. 90046	The development and implementation of a comprehensive and sequential ecology program from kindergarten through adult education has been the goal of this project. The suggested learning strategies for children emphasize a large number of individualized experiences that are each keyed to various ecological concepts.
Outdoor Biology Instructional Strategies (OBIS)	OBIS Lawrence Hall of Science University of California Berkeley, Calif. 94720	Helping children learn how to make intelligent decisions about the use of the environment is a principal focus of this project. The activities and materials of the OBIS project can be used with children aged 10 through 15. Community group leaders as well as teachers can utilize the strategies of large group, small group, and individualized instruction incorporated in the project.
People and Technology (P&T)	Education Development Center 15 Mifflin Place Cambridge, Mass. 02138	This program is appropriate for middle school children who are interested in exploring the relationship between society and technology. P&T is organized into various units which include, among other things, hands-on experiences, case studies, and a community study. Multi-media and inquiry-based learning strategies are utilized to involve children with varying abilities.
Science—A Process Approach (acronym SAPA or AAAS)	American Association for the Advancement of Science 1776 Massachusetts Ave. N.W. Washington, D.C. 20036	This science curriculum emphasizes the understanding of the natural world through a study of such processes as observing, meaning, and inferring. Scientific activities that develop and extend these processes for children from 5 to 12 years of age in a sequential manner are the basis for the project.

Science Curriculum Improvement Study (SCIS)	Lawrence Hall of Science University of California Berkeley, Calif. 94720	The acquisition of science knowledge and understanding coupled with a scientific attitude of inquisitiveness and rationality is a goal of SCIS. These combined aspects of the project are designed to help children develop scientific literacy through SCIS activities. The materials and activities are designed for use by students in grades K through 6.
Student Centered Science Program (SCSP)	San Francisco Unified School District Student Centered Science Program 350 Amber Drive, Rm. 2 San Francisco, Calif. 94131	A conceptual schemes (concepts) approach to science is the basis for this science curriculum. Students (K through 6) perform activities that are organized around the basic concepts of science. The materials for the activities are of low cost and available to teachers and children.
Unified Science & Mathematics for Elementary Schools (USMES)	Education Development Center 55 Chapel Street Newton, Mass. 02160	The integration of science and mathematics as vehicles for solving real and practical problems from the local school/community environment is the principal focus of this project. Students become involved through the application of specific problem solving strategies to real situations. The materials utilized have been developed for children from 6 to 14.
Valuing the Environment	Environmental Education Center 1658 Sterling Road Charlotte, N.C. 28209	This is an interdisciplinary approach to the environment that uses environmental packets that emphasize value clarification. In this project teachers use hands-on materials and field work to assist children (ages 6 through 11) in understanding and valuing their environment.

Webbing

An interesting planning technique that is both fun and effective is the process called "webbing." If you haven't heard of it until now, we think you might wish to try this technique that integrates the openness of brainstorming with some specification of tangible outcomes for learning.

Webbing can be done by the children at the beginning of the year or portion of the year in which they will have considerable flexibility with respect to the areas they would like to study. At least two assumptions underlie webbing:

1. The concepts, skills, and process of science are related: they are webbed.
2. That portion of the web that extends from our own interests can be drawn.

Here's how to start a web. At the center of a page or at the chalkboard, list a question that a child has about the world. A child might start a web with: *Can people live on the moon?* Ask the child to now raise related questions or statements and attach them to the original question. Encourage the child to keep adding to the web and perhaps even to begin to "branch out" with his ideas. Notice what happens as the web grows (Figure 2)?

Figure 2. Webbing.

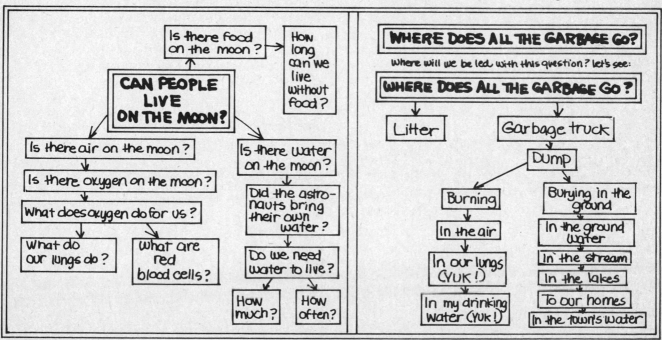

This tiny web emanating from one child's question about his or her world provides not only surprising results but an interesting road map that can guide a child's exploration. You will obviously need to assist young children in making their web. You might even be able to construct a pictorial web.

Enjoy your webbing. It's a lot more fun than trying to make up a syllabus.

Workshops

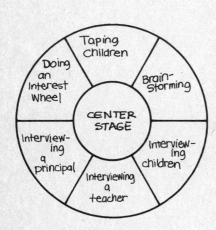

Center Stage

You have just come home to your family after spending your usual 10-hour day with your students and classes. The open learning you have been trying has been very rewarding to you personally, but the time spent has been more than you expected. Your students are pleased with the course and with having a chance to interact with you. Your family is sitting at the table, looking at you with expressionless faces. You greet them and they do not respond.

What are you feeling? What do you think your family is feeling? What do you plan to do or say about their silence? How do you plan to tell them you have to go back that night to meet with a student who needs to talk with you?[8]

Taping Children

Select two children at three different grade levels and tape record their responses to the following:

a. What do you think a scientist is like?
b. How does science affect your life?
c. Tell about people you know who work in science.
d. Would you like to work in a science-related field?

Brainstorming

Get a group of students and ask them what they would like to study in science. Brainstorm with a group of children about the question, "What topics would you like to learn about this year in science?" While brainstorming, do not evaluate student topics but rather try to generate as long a list as possible. Use a consensus technique (how many want to study this topic) to reduce the list. Did the list surprise you? Did the students have trouble coming up with topics?

Interviewing Children

Children are great fun to interview. You may want to interview children at different grade levels and ask them such questions as:

What is science?
What kinds of things would you like to study in science?

Interviewing a Teacher

Find a teacher who has more teaching experience than you have and try to get his or her opinion on:

1. Openness and accountability.
2. Science as a humanistic experience.
3. What kinds of science children at that grade level are studying.
4. How much time each week can be used for science.

Interviewing a Principal

Meet a principal who can spare a few minutes for an interview. Try to get his opinion about:

1. The emphasis science should have in the school curriculum.
2. The extent to which science should be experience based.

Doing an Interest Wheel

Draw a circle. In the center of the circle, list one thing in science that interested you as a child, then put in as spokes all the other things that an exploration of that topic could lead to.

Reference Shelf

MERRILL — FREEDOM TO LEARN — Carl R. Rogers — COLUMBUS, OHIO 1969

MERRILL — HUMANISTIC TEACHING — Donald H. Clark and Asya L. Kadis — COLUMBUS OHIO 1971

MERRILL — BECOMING A BETTER ELEMENTARY SCIENCE TEACHER — Robert Sund and Roger Bybee — COLUMBUS OHIO 1973

FAWCETT — I ain't much, Baby, but I'm all I've got — Jess Lair — GREENWICH CONN. 1974

ADDISON WESLEY — OPENING — Bob Samples and Bob Wohlford — READING MASS. 1973

ADDISON WESLEY — BORN TO WIN — Muriel James and Dorothy Jongeward — READING MASS. 1971

N.S.T.A. — PERSONALIZING SCIENCE TEACHING — Roger Bybee

MERRILL — RISK-TRUST-LOVE: Learning in a Humane Environment — William D. Romey — WASHINGTON D.C. 1974

VIKING PRESS — HUMAN TEACHING FOR HUMAN LEARNING — George I. Brown — COLUMBUS OHIO 1972

THE SCIENCE TEACHER — "SCIENCE IN THE NEW HUMANISM" J.BRONOWSKI — OCTOBER 1968 — NEW YORK 1972

PLANNING FOR A HUMANISTIC CLASSROOM

Conversation

Joe: Well, Jack, how do you feel about the chapter? Were we on target?

Jack: I'd say so. I hope everyone gets the main point. You know, the idea that helping children learn is enjoyable but very, very hard work.

Joe: Precisely, almost all the new teachers I've had a chance to work with say that they are surprised by the drain on their energy.

Jack: I've found the same thing. Before people really get a chance to roll up their sleeves, many think that a good classroom will just happen if they love kids.

Joe: I think we all go through that phase. I remember my first year—well, really the disastrous first few weeks. I thought that by just having super activities around for the kids to work on everything would just magically happen. It didn't.

Jack: Boy, does that ever sound familiar. That was my experience, too. I wish I knew then what I've learned over the years. So many things make such a big difference: taking the time to talk with the kids to see what their interests are, talking to some experienced teachers to get some science ideas, checking out new resource materials such as the "new" curriculum projects, and finding the time to do some reading in the education magazines.

Joe: All this really starts to look like so much work, especially for someone new to teaching. I hope we don't discourage people because the rewards of working with kids are really great. But where will they find the time to do all these things?

Jack: I wish I knew the answer to that.

Pastel and India ink.

2
Fostering Student Involvement

Children are different. They differ from each other emotionally, physically, intellectually, and socially. They look at themselves and their world through different pairs of eyes.

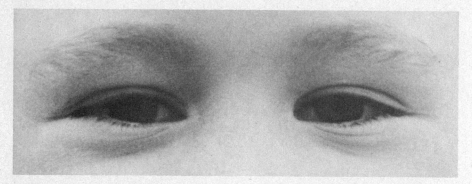

We must reflect upon the range of perceptions children have to fathom truly the enormity of the challenge we face in the classroom. How can we provide a humanistic learning environment for the range of children that come to us each day?

As a result of our work with many different teachers at a variety of grade levels, we feel that any teacher committed to providing a meaningful response to the challenge described above must establish a science learning environment that includes a variety of learning experiences. Three types of learning experiences would seem to be at the core of a successful science program:

1. Individual learning experiences
2. Small group learning experiences
3. Large group learning experiences

Individual Learning

Throughout the year each child needs opportunities to work on his own science activities. Many science activities lend themselves to independent investigation and provide a chance for the child to become more self-reliant. One important goal for us as teachers is that of fostering self-confidence in each of our girls and boys. Each child needs to learn that he or she is fully capable of initiating a science activity and carrying it out.

Happiness is discovering something on your own.

Cultivation of this sense of responsibility for one's own learning will benefit the child not only in science but in all other areas of the curriculum as well. Children who develop "self-responsibility" will have the power to achieve success in schools and in the outside world irrespective of the strengths and weaknesses of the academic or social environment in which they find themselves. Yes, a responsibility for their own learning has enormous meaning and importance to the lives of children.

We can consider our teaching strategies for fostering student independence and initiative by taking a close look at two components of individual student investigation:

1. Choice making
2. Action taking

Choice Making

One of the overriding characteristics of humanistic classrooms is the responsibility placed on the child for choosing what he will investigate. To some extent, philosophical and curricular constraints unique to your particular school will provide some boundaries to a child's choice making. We hope, however, that in your classroom you will encourage as much choice making as is realistically possible.

The importance of choice making, even if it is only between two different activities, cannot be overemphasized. Each girl and boy comes to us with a certain curiosity about the world and a set of perceptions about that world. When given a choice, the child will pick the activity that has the most importance to him, thus increasing the likelihood of a successful science experience.

A gastronomic example might make this point more clearly. Let's say you are going to take a group of friends out for dinner and there are two restaurants that each have an excellent service and an excellent chef. There is, however, one important difference, the menus.

	Chef Gabriella's Menu		*Chef Pierre's Menu*
Appetizer	Shrimp Cocktail	**Appetizer**	Choice of: Shrimp Cocktail Vichyssoise Consommé
Salad	Lettuce and Tomato with French dressing	**Salad**	Choice of: Lettuce and Tomato with French, Russian, Italian or Blue Cheese dressing Fresh Fruit Salad
Entrée	Prime Ribs of Beef	**Entrée**	Choice of: Prime Rib of Beef Coq au Vin Coquilles St. Jacques
Dessert	Creme de Menthe Parfait	**Dessert**	Choice of: Creme de Menthe Parfait Baked Alaska Cheese Cake with Strawberries

Although both restaurants serve excellent food, a dinner party at Chef Pierre's is going to be more successful. Each member of the party can select those foods he or she prefers, and the chances of good meals for everyone are increased. The menu situation is analogous to providing choices of science activities for children. Our menu might be limited for many reasons, but we must try our very best to provide at least some student selection of activities. In the second part of this book, a variety of activities that foster discovering, valuing, and exploring are presented. We hope that they will assist you in providing good choices for children.

Action Taking

The child who has made a choice for an investigation must then follow through and carry out the work necessary to reach a conclusion. To develop the sense of excitement and accomplishment that comes with carrying out a science experience, the child must feel that he or she is the one who is successful, not the teacher. As teachers we sometimes have a tendency to hover over a child's work, thereby making sure that everything is done "right." Development of a positive self-concept, an important humanistic goal for the science classroom, is very difficult if the child feels the teacher is the person who really did the work. The child needs to know that we are nearby if we are needed, but not so close as to suppress his independence of action.

By letting the child take his own action we let him earn his success and learn from his failures. The freedom to learn from successes and failures offers children an opportunity to practice the self-reliance so desperately needed for a happy, productive life. When we do too much for children, we inadvertently work against the development of a healthy self-concept and give children a learning and social problem they may never overcome.

Small Group Learning

Forming Science Learning Groups

"Bleckgh, I've got to work with him!? Yuk!!"

"Aghh, I've got to work with her!? Yuk!!"

One overriding purpose for science group work is presenting an opportunity for children to work cooperatively. After all, the spirit of humanism is the spirit of respect and faith in others and in the belief that, by working together, many things become possible. The children we work with deserve many opportunities to work together to learn about themselves—and to learn about science.

Science learning groups reflect in their composition the range of abilities and personalities children are likely to encounter for the remainder of their schooling and in adulthood. This fact poses for us a rather difficult challenge: how should these learning groups be formed? *Should we let children form their own groups or should we organize the groups?* The answer, of course, is found if we reflect for a second upon the nature of the groups the child will be likely to find himself in as he goes through life.

The child is part of a family group, may be part of a play group (neighborhood children whom he or she sees after school), a clique of peers who have been together in previous years, church or synagogue groups, scouting groups, and recreational groups. As an adult, she or he may become part of various civic and social groups. The composition of these groups varies from

Working together inside and outside the classroom.

those that are very heterogeneous to those that are quite homogeneous. The growing child and the fully functioning adult must develop the group process skills that will allow him or her to function at maximum potential with a variety of persons.

Of course, people who think that they don't have a lot in common may find that they really do share more than they would initially think. If you've read *I'm OK, You're OK*, you may remember that Harris notes that "*persons are important* in that they are all bound together in a universal relatedness which transcends their own personal experience."[1]

Based on all of the above, the conclusions for us as teachers with respect to the formation of learning groups seem clearer. Children in science as well as other subject areas need a *variety* of small group experiences. Somehow we must arrange opportunities for children to form their own groups freely and yet we must exercise leadership to ensure that children of disparate backgrounds and abilities have a chance to work and grow together.

There is of course no fixed number of small group experiences of each type that all children should have. Just try your best to intersperse self-selected groups with planned heterogeneous groups as the year proceeds. By learning to work with people who are like ourselves and people who are different, we become more aware of our own humanity and become better equipped philosophically to put our concern for our fellowman into practice.

Children may feel that they face the challenges of life and schooling on their own. As adults we may also feel this way from time to time. The lyrics of this song quite beautifully express a contradiction to the feeling of loneliness.

EVERYTHING THAT TOUCHES ME

You say you're all alone
The vision is all your own
The things you feel
Nobody feels the same
But you know me
You know what I've been through
And everything that touches me
And everything that touches me
And everything that touches me
Touches you.[2]

Nothing brings forth our capacity for humanistic interaction as well as an ample opportunity to interact with others. One of the central purposes of schooling is to assist children in developing socially: cooperative science activities can provide the medium in which that social growth can occur.

When two children work together on a science activity, they experience science and they experience each other. Both experiences are crucial to the development of adults who understand and appreciate science as a humanistic endeavor. Science is more than facts, concepts, skills, and processes: it is an activitiy through which humans working together unlock the secrets of the natural world.

In the classroom setting the study of science is a very natural way for you to assist children in developing the skills required for productive, cooperative efforts with their peers. Group science activities provide a chance for them to share their thoughts, their skills, and their interests. They provide an opportunity for children to learn that sometimes two or more people working cooperatively can learn more than one person working alone—and perhaps enjoy it more.

Of course, there are problems you should be sensitive to if you encourage group science activities. These include:

1. Tarzan-Jane problems
2. Too many cooks
3. The social butterflies

Small Group Pitfalls

Tarzan and Jane

Sexism in the science classroom emerges most clearly during group work. Somehow, almost magically, a boy and girl who are supposed to be working cooperatively on an investigation become transformed into a tacky "Me Tarzan, you Jane" sketch. There seems to be something about the way in which scientific inquiry is reported to our citizenry that identifies the scientist role as a masculine one. As young children watch TV, listen to adult conversation, and generally learn about occupations, they begin to stereotype the masculine scientist role. Boys see themselves as potential scientists, doctors, and engineers, and girls see themselves as potential helpers. Group work in the science classroom unfortunately provides an opportunity for boys and girls to play out the stereotypes in the format of scientist and secretary. The boy becomes the "doer," and the girl becomes the recorder of the observations and the "gopher." ("Mary, gopher [go for] the string we need.")

As teachers we have a responsibility to the boys and to the girls to help them all reach their maximum potential. The sexist role traps they have learned to believe in are comfortable, and it will take our conscious, overt effort to help children see through these sexist constraints. The implications are clear: in our behavior we must show the scientific role to be one that can be filled by females as well as males; we must learn to say, "The scientist was surprised by her discovery." Our behavior during the formation of groups must encourage boys and girls to work together. During the course of experimentation, we must watch carefully for any division of labor according to Tarzan and Jane roles. When we see such roles being played within a science group, we must suggest strongly that all parties share project responsibilities equally.

Too Many Cooks

Sometimes science groups grow too large for successful exploration. Mysteriously a group of three or four starts to grow to five or six, and perhaps even more. You may find this particularly true with older children as they seek to keep social cliques together. For most science activities, a two- or three-person group will be adequate. If the groups become much larger, the old adage about "everyone's responsibility being no one's responsibility" begins to come true.

How do you feel about the size of this group?

As group size increases, we notice the development of "peripheral people." Peripheral people are those boys and girls who are on the outer fringes of the activity. They may "ooh and aah" as discoveries are made, but they seldom have a chance to roll up their sleeves and become involved. When you observe too many "peripheral people" during the experiments, you will need to intervene and form additional groups.

The Social Butterflies

Group work in science has two purposes: investigating science, and social development (cooperativeness). Sometimes in our work with children in science we may notice that some children may fail to keep both dimensions of group work in balance. The environment we have in our classrooms during science time should be one that encourages lots of interaction among group members and relatively less interaction across groups. It is certainly appropriate for children to stop by other groups for a brief visit; however, this can be carried to an extreme. Some children may see science group work as an opportunity to socialize with everyone else in the classroom. These social butterflies may need some occasional private reminders about trying to channel their energies within their own group.

One way of providing a productive outlet for this desire to socialize with other groups is to take 5 or 10 minutes at the end of a science period before each group "cleans up" for a sharing time. Children could then visit with other groups to socialize and learn what others have been working on.

Group work sometimes puts us in the position of deciding as the instructional leader whether a particular set of behaviors, such as those exhibited by "social butterflies," is appropriate or not. Making the decision is difficult because we wish to foster both science learning and social development. The judgment we make will reflect our personal philosophy of science education and the needs of children. The balance between freedom and license is a difficult one to conceptualize and even more difficult to foster in our classrooms.

Small Group Openers

Here are several small group activities that we have found very helpful in dealing with the small group problems just discussed. The value of these activities lies in helping you and your students come to grips with small group process skills, not in the content that may emerge from them.

Disaster on the Moon

Assign each student to a group no larger than three or four. Have members of each group sit together on the floor or arrange their chairs in a circle. Now read or tell the students this story:

> You are a member of moon space crew originally scheduled to rendezvous with a mother ship on the lighted surface of the moon. Due to mechanical difficulties, however, your ship was forced to land at a spot some 200 miles from the rendezvous point. During reentry and landing, much of the equipment aboard was damaged; and because survival depends on reaching the mother ship, the most critical items available must be chosen for the 200-mile trip. On this sheet of paper are listed the 15 items left intact and undamaged after landing. Your task is to rank order them in terms of their importance for your crew in allowing them to reach the rendezvous point. Place the number 1 by the most important, and so on through 15, the least important.

After each student has a list (Table 3), tell the students that their group is to employ a method of group consensus in reaching a decision about the rank ordering of each item. Let each group decide upon its own procedure, but do indicate that group members should:

1. *avoid arguing for individual judgments;*
2. *avoid voting or averaging in reaching decisions;*
3. *view differences of opinion as being helpful rather than as a hinderance to group process.*

After no more than 30 minutes, have members of each group report the *process* they used in arriving at their decisions and read their list of rank ordered items. Having the list on a large sheet of paper or on the chalkboard will help. Finally, share the list of NASA's choices with your students.

Table 3.	
"Disaster on the Moon" Lists	
List 1: Undamaged Items (Give this list to the students.)	**List 2: NASA's Choices and Reasons** (Hold this list for later in activity.)
_____ Box of matches	_15_ Little or no use on moon
_____ Food concentrates	_4_ Supply daily food requirement
_____ 50 feet of nylon rope	_6_ Useful in tying injured together, help in climbing
_____ Parachute silk	_8_ Shelter against sun's rays
_____ Portable heating unit	_13_ Useful only if party landed on dark side.
_____ Two 45-calibre pistols	_11_ Self-propulsion devices could be made from them
_____ One case dehydrated milk	_12_ Food, mixed with water for drinking
_____ Two 100-lb. tanks of oxygen	_1_ Fills respiration requirement
_____ Stellar map (of moon's constellations)	_3_ One of principal means of finding directions
_____ Life raft	_9_ CO_2 bottles for self-propulsion across chasms, etc.
_____ Magnetic compass	_14_ Probably no magnetized poles; thus, useless
_____ 5 gallons of water	_2_ Replenishes loss by sweating, etc.
_____ Signal flares	_10_ Distress call when line of sight possible
_____ First aid kit containing injection needles	_7_ Oral pills of injection medicine valuable
_____ Solar-powered FM receiver-transmitter	_5_ Distress signal transmitter possible communication with mother ship

Five Squares[3]

Another opener we suggest is one that is a little different from Disaster on the Moon. It is called Five Squares and is a nonverbal group activity. That's right, no talking. (If you think it's difficult trying to get your kids not to talk, try it with a group of teachers!)

This game is a fantastic way to get at the process of group cooperation in solving a problem. To play the game, let the students form groups of five sitting around a table or on the floor. Prepare the five squares by cutting shapes out of cardboard using the patterns shown in Figure 3. Place the five square shapes in envelopes using the groupings shown in Figure 4. These groupings are for one group, so you'll have to prepare enough for each group. (Label each piece as shown in Figure 4 to avoid postactivity disaster.) Give the students these directions:

Each of you will be given an envelope. The purpose of the game is to form five squares such that each player in your group has a square the same size. During the game you may not talk. You must maintain complete silence. The only thing you may do is give a piece of your puzzle to another player. You must hand it to the person you wish to give it to. You may not point to a piece to indicate that you want it, nor can you put a piece you are giving away in another person's square. The game *is over when each person has a square in front of him the same size as everyone else's. You have as much time as you need to solve the problem.*

If you have one or two students left over because of groupings based on five, have them participate in the activity as observers. Give each a card with instructions to look for some of the following behaviors in the groups: cooperation, laughter, anxiety, frustration, patience, indifference, anger.

When all the groups have solved the problem (which may take from 10 to 40 minutes), arrange everyone in a circle and discuss the activity. Making use of the observers as well as the players, focus on such questions as:

How did you feel about playing the game?

Did you feel a sense of cooperation among members of your group?

Was anyone selfish in your group? How?

Did boys and girls participate in similar ways? How?

How did you feel about not being able to talk?

Did you feel anger or hostility toward group members?

How did you feel about having other groups finish before your group?

How do you like working in a group as opposed to working alone?

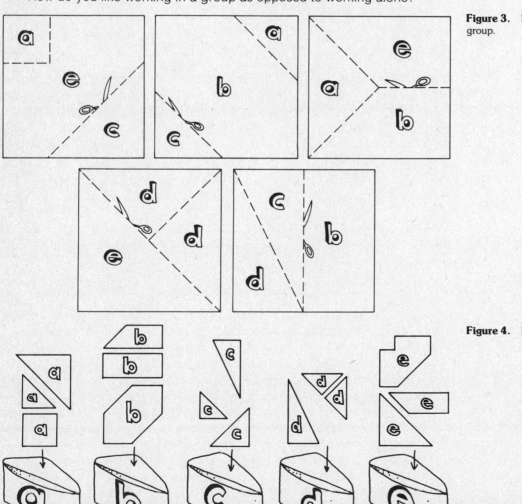

Figure 3. Five Square shapes for each group.

Figure 4. Five Square groupings.

Large Group Learning

Class Experiment

One aspect of humanistic science education is the cultivation in children of a vision of the classroom as a community. Our hope is that this sense of community within the school setting will provide a basis for active civic participation in their adult lives. At various times in the school year we can begin this process of community building by having the class try a few large group explorations.

The classroom as a community: Discussing alternative learning activities as a large group in a class meeting.

Each child or group of children can work on separate but related components of the investigation and contribute his work and products to the overall exploration.

You can assist the children by being responsible for placing a list of alternatives on the chalkboard and getting them to add to the list. The next step is the decision-making process through which the class selects a class investigation. The listing of the components and possible actions will prove to be an excellent discussion activity for the class. Finally, you will need to encourage the children (or groups of children) to select a component for exploration.

The scope of possible large group explorations is large; however, the time and resources available will affect the lists generated by you and your students to some extent. You might find the examples in Table 4 helpful in organizing class planning.

When all the individual and group investigations are complete, the individuals and groups need an opportunity to report back to the full class so that all

Table 4.

The Classroom as Community—Possible Explorations

Exploration	Components	Actions
Adopt a stream	Plants in the stream	How many kinds?
	Plants on the stream bank	How many:
	Dead things in the stream	Leaves, twigs, fish, frogs,
	Dead things on the stream bank	insects?
	Water smells / tastes	Describe them.
	Rocks in the stream	How many kinds?
	Rocks on the bank	Are they big, small smooth, rough?
	Water speed	Time a floating twig.
	Pollution	Count man-made things— beer cans, trash, etc.
Conserving water in school	Drinking fountain	How much water goes down the drain every time you take a drink?
	Bathroom	Are there broken toilets that keep running?
	Kitchen	Do the cooks waste water when they are getting the food ready?
Weather study	Weather chart	Make a chart.
	Temperature	Take a temperature and record on a chart.
	Cloud cover	Draw the types of clouds and record on a chart.
	Wind direction	Use a compass and record on a chart.
	Air pressure	Read a barometer and record on a chart.
	Weather maps	Study newspaper map. Paste on chart.
	Weather forecasts	Call local weather station.

can profit. The atmosphere should be one in which each child feels that he has made an important contribution to the larger class effort.

Show and Tell, or Share and Tell, or Bring and Brag

Children love to play out adult roles: young children like to play mommy, daddy, doctor, nurse, fireman. One adult role that can give them a special feeling of leadership is that of teacher. In science we can find some times during the year when children who wish to can do an experiment in front of the entire class, tell about a nature walk, or share other science experiences. It's a good feeling being in front of a class and being the focus of everyone's attention. "Show and tell" time is one of the names that teachers use to describe this time. We feel the term is a little babyish. Other teachers use the term "share and tell," which has a much nicer ring to it. In jest some teachers even call it "bring and brag" time!

If the child showing is the "bragger," are
these children the "braggees"?

Share and tell time for science provides children with a chance for a leadership experience that uses science as a medium. Keep the following things in
mind if you try some science "share and tell":
1.   Participation should be voluntary.
2.   Any one student's time should be limited to 5 minutes so that all
     interested children can participate.
3.   Don't overdo it. "Share and tell" days are special days, and the
     children may lose interest if you have too many of them.

Teacher Demonstration

There are times during the year when you will feel the need to take center stage
and present a science demonstration for the class, a time for you, the teacher,
to be the focus of everyone's attention as you share an important concept with
them, pose some alternative activities, or generally try to pique their curiosity
about something they may not yet have explored. The demonstration is an opportunity for all children to experience a scientific phenomenon that, for one
reason or another, they would not have an opportunity to experience individually or in small groups.

The demonstration is not simply an excuse to present a long lecture to children. Rather it is a chance to help the students see or hear something that will
help them get "charged up" over science. Experienced teachers become
gatherers of demonstrations that work well with children. You may wish to begin a collection of demonstrations that you can draw on as the year goes by.
Simply place the necessary information on file cards as in Figure 5 and use the
demonstration as a resource. Here are two that you may wish to use to get your
collection started. Have fun.

Sound

Have the children shut their eyes while you conduct this demonstration. You
will have to acquire the following sound-producing objects before class and

Figure 5.   Sample demonstration file
cards.

SOUND

Have kids close eyes and guess.

1. Plastic police whistle
2. Toy drum
3. Sandpaper & wood
4. Tuning fork
5. Soda bottles...fill at different hts.
   (either blow or tap with pencil)
6. Rubber bands...diff. thicknesses......
   on cigar box

LIGHT

Darken room and put up
screen or poster.

1. Three identical flashlights
2. Red, blue, green cellophane
3. Three rubber bands

(color "addition" and "subtraction")

hide them in your desk:
1. Plastic "police" whistle;
2. Toy drum;
3. Sandpaper and a piece of wood.
4. Tuning fork;
5. Soda bottles partly filled with various amounts of water (make the sounds by blowing across the mouths or tapping with a pencil);
6. Rubber bands of different thicknesses wrapped around a cigar box.

Ask the children to shut their eyes before you make each sound. Take one of the objects from your desk and produce the sounds. Ask the children to guess what you are using to make the sounds. After they guess, have them open their eyes. This demonstration can serve as a nice motivating technique to get children started on an exploration of sound.

Primary Colored Lights

For this demonstration you will need a room you can darken. Good shades, curtains, or windows that can be covered over with newspaper will provide the light blockage you will need. For the demonstration you will also need:
1. Three flashlights having the same number of batteries and same lens size;
2. Red, dark blue, and green cellophane (you can buy them in a stationery store);
3. Three rubber bands;
4. A projector screen or some poster size pieces of white paper taped to your blackboard.

Using the flashlights, cellophane, and rubber bands, make each of the flashlights a different colored spotlight: one red, one blue, and one green. During the demonstration, have three children come to the front of the room to shine the lights and discover what happens when various combinations of the primary colored lights are mixed. (If you want to learn more about this phenomenon, you can use almost any junior high school or high school physical science textbook as your personal source book: look up "color addition." (Mixing primary color pigments together results in a quite different phenomenon called "color subtraction," which would also make an interesting demonstration—but we'll let you think that one up on your own.)

A successful teacher demonstration is one that is
1. brief,
2. visible (smellable),
3. apt to lead children to follow up the demonstration with their own work.

Demonstrations are a great deal of fun for the children, and for you. Sometimes they can be disastrous, particularly if you don't try them out before class to make sure you have every piece of material you need.

As we have said many times throughout this book, we must be careful to avoid overdoing things. Teacher demonstrations should never, ever, ever take the place of having the children work with the "things and stuff" of science. Look at the demonstration as an opportunity to add a little extra seasoning to science class because

a little spice

is

nice.

Workshops

Center Stage

You have decided to devote the month of May to individual science investigations with your fourth grade students. You've told the students that they are free to decide on a project or group of projects for investigation, and that they will be required to assess themselves on their effort and accomplishment.

It is now the end of May and you've asked the students to report to you on their accomplishments. Mark, an eleven-year-old who has had a number of problems in the earlier grades, tells you that his major individual accomplishment, the growing of three bean seedlings under three different light conditions, was a disaster because the custodian put the seedlings on the heating unit one night and they died. He (Mark) threw them out yesterday, lost his notebook, and now has no evidence that he really did a project.

1. What would you say to Mark?
2. What grade would you give him for the marking period?
3. What would you say to the custodian?

Large Group Learning Experiences—Good

Have you ever had a good large group learning experience? What were some things that made it a good experience? Find someone else who has had a good large group experience. Share the good things that happened. Think about

1. the leader (facilitator)
2. the topic or project
3. the amount of work
4. your role

Large Group Learning Experiences—Bad

Have you ever had a bad large group learning experience? What do you remember about it? Check to see if anyone else you know has had a bad experience. Compare notes about

1. the leader (facilitator)
2. the topic or project
3. the amount of work
4. your role

Figure 6. Personal shield.

Textbook Topics

Find a science textbook at any grade level. Pick a chapter you think would interest children. Then list the topics or headings under the type of learning experience the topic would be best suited for:

1. individualized learning
2. small group learning
3. large group learning

Personal Shield

Make a drawing of a shield with four blocks (see Figure 6). Number the blocks. In block 1 write something in science you would enjoy working on by yourself; in block 2 list something you would like to explore with a friend; in block 3 list something you would like someone to tell you about; and in block 4 with a pie diagram show the relative amounts of individual, small group, and large group instruction you feel should take place in a humanistic classroom. Compare your shield with shields others have made.

Opinion Poll 1

Can a teacher doing large group instruction be humanistic? What do you think? Compare your opinion with someone else's.

Opinion Poll 2

Can individualized instruction be nonhumanistic? Compare your opinion with someone else's.

Reference Shelf

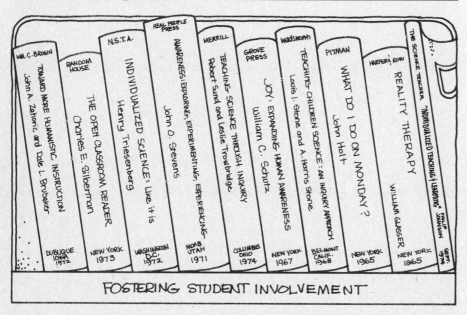

FOSTERING STUDENT INVOLVEMENT

Conversation

Jack: I liked the way we started out by emphasizing the eyes of a child. The message there really keeps me on my toes as I think about teaching and learning.

Joe: I think I know what you mean. Even though I think we have such good intentions as teachers—you know, trying to individualize as much as possible, doing some small group work, and some large group work—I sometimes forget that each person I try to teach is looking at me and the rest of the world in a different way.

Jack: We get so involved in doing things that it really is easy to forget to look at the world the way a child might look at it.

Joe: I think I can recall something that happened with a group of kids I was teaching that really brings this point home. I was doing some large group instruction about the difference between Centigrade and Fahrenheit temperatures, and about midway through I decided I was going to show the class on the blackboard a quick way of changing Fahrenheit temperatures to Centigrade. So I started to explain it. I was really excited about it. So I worked at the board for a few minutes, and I didn't pay too much attention to the kids seated by the window. When I looked over toward them, there they were, six kids all intently staring out the window, watching a steam shovel that was excavating the hole for the basement of a new Board of Education building.

Jack: How did you feel?

Joe: Crummy.

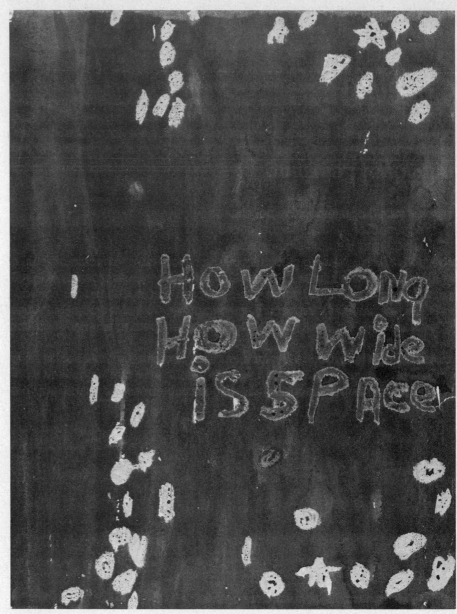

Crayon and paint (crayon resist).

3
Acquiring and Using Resources

T he science classroom should be a place rich in stimuli that will reach out and touch the minds of children. The classroom should be for them. A conventional classroom that is sterile except for an old bulletin board display and an obscure, dust-covered "science corner" simply has little effect on youngsters brought up in our dynamic, high-tempo world. The child who is stepping to the beat of our rock-and-roll world outside comes to school and finds that his teachers are playing a waltz.

For better or for worse, children need a rich, out-reaching environment to catch their attention. We must try to develop such an environment in our classroom. To accomplish this goal we will need to put away some of our old-fashioned thinking about resources.

Here is a classroom in which the teachers and children have created lots of learning opportunities by using their resources creatively. Would you like to take a closer look?

Floor: Here is a floor game (giant Tic-Tac-Toe) which is played by answering questions printed on cards. One team gets a supply of paper X's while the other team is given O's. They get to use them each time they answer a question.

Using Resources

The science curriculum should have as its principal focus student exploration and experimentation. As the child begins to investigate the natural world, various questions related to his experimentation will emerge. In the conventional classroom the teacher, a group of textbooks, or a set of encyclopedias has been the student's principal resource. As we move to a more open classroom environment, we need to develop resources that more closely correspond to the type of curriculum that stresses student responsibility and activity.

The teacher's role in the classroom is one that requires him or her to resist the temptation of simply answering every question the child raises. The teacher must be a "facilitator" who redirects the child and helps the child discover the information he needs through the use of readily available resource materials.

Severe criticism can be leveled at teachers who merely direct a child to "go look it up in the encyclopedia." This might be somewhat effective with very

Wall: Up, up, and away! Using a wall display to study flight.

Corner: The children and teacher worked together to develop this science corner. Most of the materials here were brought to school by students.

bright children who are able to master the process of using the encyclopedia index systems and who have sufficient reading skills to secure the information from the printed page. Many children will probably find encyclopedias beyond their reading skill, dull, and frustrating to their healthy curiosity.

Another resource that is largely misused is the school library. Many times students having questions are simply directed to the library. Teachers sometimes presume that sending a child to the library is, in and of itself an educational experience for the child. For many children the library is a rather confusing place, and the seemingly endless process of digging out information may overwhelm the young child. A competent school librarian can certainly facilitate this process; however, she is also responsible for handling *all* the children who arrive at the library's portals. Very little time is available for her to work with children on an individualized basis.

As you prepare to teach science in a humanistic classroom environment, you will need to view the utilization of resource materials within a larger context. Certainly encyclopedias and libraries will play a part in the science education for the child, but they should do just that and not become the entire resource system. There are many other resources that can be developed.

Minibooks

Giants design textbooks, or so it must seem to children: they are big, heavy, and clumsy. The reason third or fourth graders have little interest in taking books home may have more to do with the inconvenience of doing so than with lack of interest in the materials. As you look through the pages of brand new textbooks, you may notice beautiful photographs and line drawings, and generally high quality graphics, all designed to pique the interest of children. Unfortunately such attractiveness is not quite strong enough to overcome the negative impressions students have about text materials.

Aside from their bulk, texts are paired in a child's mind with ''punishers.'' The textbook is thought of by children as a compilation of complex reading selections. How can we as teachers overcome this feeling of despair that can arise when we wish to use a science text as a resource? Using a text as a resource is a perfectly acceptable function within a classroom that fosters intellectual curiosity, but it must be used properly. Don't be fooled by those who would say that a text is a ''no, no.'' As with everything else in life, the value of something depends on what you do with it. Any given textbook can be a ''punisher'' or a ''goody,'' depending on you.

One approach to making a textbook into a useful resource is the construction of minibooks. A minibook can be one of two types:

1. A portion of a commercial book that has been removed and rebound for a special purpose.
2. A small book prepared by students for use by other students.

We will deal with each of these types of minibooks.

Rebound Commercial Book

To prepare the first type of minibook mentioned above, you will need to assemble a small collection of science texts from different grade levels. Next, remove all the pages from the binding by using a single-edged razor blade to slit the paper glued inside the front and back cover of the book near the binding. Then make a minibook of each chapter or part of a chapter that deals with a particular topic. You or the students may make a front and back cover for each minibook of cardboard and then cover the cardboard with "contact" paper.

How can we use minibooks in our teaching? Take the example of a teacher who is preparing resource materials for the major theme "From Seeds to Plants." By searching around the cupboards and closets of the school, the teacher may locate science texts from a variety of grade levels. A few texts from different grade levels will provide an abundance of minibooks. Looking through each book, the teacher identifies all chapters that deal with some aspect of the general theme "From Seeds to Plants." For example:

Chapters	Pages
Series A	
Book 3—"Starting a Garden"	pp. 17–23
Book 4—"What Are Seeds"	pp. 42–50
Book 5—"How Plants Grow"	pp. 36–46
Series B	
Book 2—"Plants Around Us"	pp. 10–14
Book 3—"Farming"	pp. 29–45
Book 4—"A Garden for You"	pp. 36–43
Book 5—"Plants—Tiny Food factories"	pp. 75–83
Series C	
Book 1—"Leaves"	p. 9
Book 2—"Seeds"	pp. 12–15
Book 3—"Fruits"	pp. 36–40
Book 4—"Indoor Gardens"	pp. 50–60
Book 5—"Helping Plants Grow"	pp. 45–53
Book 6—"Fruits and Flowers"	pp. 75–85

Each selection becomes a minibook. The student now has an excellent collection of resource materials. If you use the minibook approach, your classroom becomes a place that has materials at a wide range of grade levels and your task of helping a child find something he is interested in and able to read becomes easier.

Student-Developed Minibooks

The second approach to helping children learn with minibooks gives the child an opportunity to prepare a minibook others can use. All you need to do as the teacher is provide quantities of blank paper cut into 6" x 7" sheets or whatever size seems appropriate. As the child does experiments or conducts activities, he records his observations on the small sheets of paper. Crayon drawings, small paintings, and pictures from magazines can all be used as illustrations. All this can be compiled by the child into a minibook that he binds and titles. All minibooks prepared by the students can be shared as parts of a classroom library of minibooks on science interest areas. As you can see, the minibook approach is a marvelous opportunity to integrate science, reading, writing, and art in a creative way.

Getting Free and Inexpensive Materials— Science on a Shoestring

When teachers talk to administrators, visions of dollar signs dance in both their heads. Teachers and administrators, for one reason or another, think that teaching science will cost a fortune. This is definitely not so, particularly when we focus upon activities that will serve to stimulate curiosity about the natural world in which the child lives. The child's world does not consist of elaborate, complicated materials and equipment; consequently, science experiences should be carried out as simply as possible.

One way of gathering materials is simply to collect those things that your children will need and store them in a central area in your classroom. Such things as rulers, string, aluminum foil, jars, cans, and rubber bands are readily available in almost every school. Make friends with the cooks, custodial staff, carpenters, plumbers, and electricians who spend time in your building. They are *important* people. They are the people who have access to many of the things you are going to need.

Other great sources of materials are the storekeepers, professional persons, tradesmen, and owners of manufacturing plants in your community. All of these people, if approached properly, will try to be helpful to you. One way to inform community members of your needs is to send them a ''scrounge list'' and cover letter. The scrounge list tells them specifically what things you are trying to accumulate.

We've included a sample scrounge list and letter (Figure 7) that have been used successfully over the years by Douglas Varney, an experienced middle school teacher. You may wish to model yours after his, or to modify the letter and use it to contact the parents of your students. It would be wise to check with your principal before doing so. Most will probably say it's appropriate to send out; some may be concerned because parents may infer that the school is not supporting science adequately (which may, of course, be true!).

Figure 7. Sample scrounge list and cover letter.

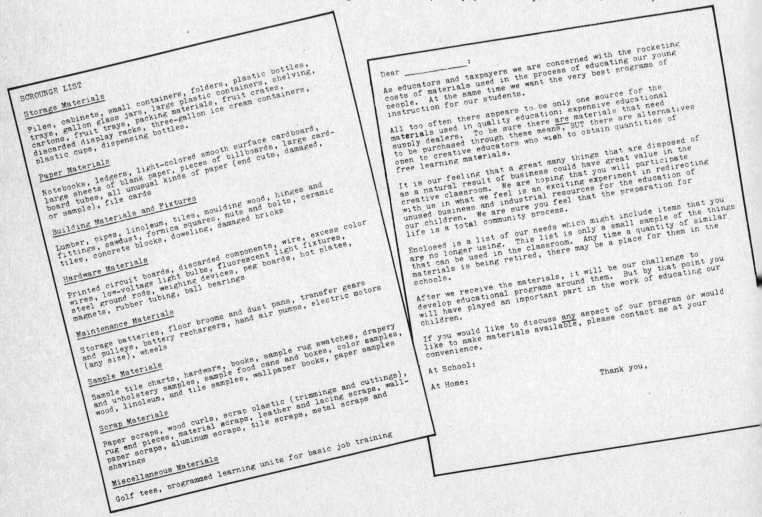

Many associations and organizations will send you free or inexpensive printed materials, such as pamphlets, brochures, booklets, and charts. To find out about the availability of such materials, simply prepare letters on your *school stationery* requesting copies of free materials or lists of inexpensive materials that are available. A list of associations and organizations you may wish to contact is shown in Figure 8.

Figure 8. Sources of free or inexpensive materials.

American Association for Health,
  Physical Education, and
  Recreation
1202 16th Street, N.W.
Washington, D.C. 20036

American Fisheries Society
1040 Washington Building
Washington, D.C. 20005

American Forest Products
  Industries, Inc.
1835 K Street, N.W.
Washington, D.C. 20006

American Forestry Association
919 Seventeenth Street, N.W.
Washington, D.C. 20006

Animal Welfare Institute
P.O. Box 3492
Grand Central Station
New York, New York 10017

Arizona Game and Fish Department
1688 W. Adams Street
Phoenix, Arizona 85007

Bureau of Indian Affairs
U.S. Department of Interior
Washington, D.C. 20242

Conservation Foundation
1250 Connecticut Avenue, N.W.
Washington, D.C. 20036

Environmental Defense Fund
P.O. Drawer 740
Stony Brook, New York 11790

Fish and Wildlife Service
U.S. Department of Interior
Washington, D.C. 20240

Fish and Wildlife Service
Bureau of Sport Fisheries and
  Wildlife
U.S. Department of Interior
Short Course Building
Rutgers State University
New Brunswick, N.J. 08901

Friends of the Earth
30 East 42nd Street
New York, New York 10017

Georgia Game and Fish Commission
Education and Information
  Division
401 State Capital Building
Atlanta, Georgia 30334

Information Office
U.S. Department of Agriculture
Washington, D.C. 20250

Izaak Walton League
1326 Waukegan Road
Glenview, Illinois 60025

National Audubon Society
1130 Fifth Avenue
New York, New York 10028

National Geographic Society
17th and M Streets, N.W.
Washington, D.C. 20036

National Parks Association
Washington, D.C. 20009

National Wildlife Federation
1412 Sixteenth Street, N.W.
Washington, D.C. 20036

Sierra Club
1050 Mills Tower
San Francisco, California 94104

State Conservationist of the
  Soil Conservation Service
U.S. Department of Agriculture
Washington, D.C. 20250

State Department of Conservation
Conservation Education
1516 9th Street
Sacramento, California 95814

Tuberculosis and Respiratory Disease
  Association
National Headquarters
1740 Broadway
New York, New York 10019

U.S. Chamber of Commerce
National Resources Department
1614 H Street
Washington, D.C. 20006

The Wilderness Society
729 Fifteenth Street, N.W.
Washington, D.C. 20005

Wish Books—Science Catalogs

If you do have a modest budget for science materials, you will need to spend it effectively and get the most for your money. Catalog shopping will help you see the types of science materials and equipment available and their costs through the various companies. Major science education suppliers are listed in Figure 9. A one-sentence letter on *school stationery* to each company requesting its catalog will bring loads of catalogs to your mailbox.

Figure 9. Wish books.

Carolina Biological Supply Co.
Burlington, North Carolina 27215

Carolina Biological Supply Co.
Gladstone, Oregon 97027

CENCO
2600 S. Kostner Avenue
Chicago, Illinois 60623

Edmund Scientific Co.
Barrington, New Jersey 08007

Lab-Aids, Inc.
Cold Spring Harbor
Long Island, New York 11100

LaPine Scientific Co.
6001 South Knox Avenue
Chicago, Illinois 60629

MacAlaster Scientific Co.
60 Arsonal Street
Watertown, Massachusetts 02172

E. H. Sargeant and Co.
4647 West Foster Avenue
Chicago, Illinois 60630

Science Kit, Inc.
2299 Military Road
Tonawanda, New York 14140

Stansi Scientific Co.
1231 North Honore Street
Chicago, Illinois 60622

TURTOX/CAMBOSCO
8200 South Hoyne Avenue
Chicago, Illinois
or   342 Western Avenue
     Boston, Massachusetts 02135

WARD'S
P.O. Box 1712
Rochester, New York 14603
or   P.O. Box 1749
     Monterey, California 93940

Welch Scientific Co.
7300 North Linder Avenue
Skokie, Illinois 60076

Making Your Own Science Films

Have you thought of making films as an activity for science class? The technology of motion picture film making has become so simplified that any child, even a six- or seven-year-old, can make his own films. The breakthrough in technology that made it possible for anyone to shoot brilliant color motion picture film was the development of the Super-8 film cartridge.

Perhaps you've used a Super-8 film camera and projector already. If you haven't, borrow them and try your hand at film making: it's really simple. The medium of Super-8 film and the large number of people who own cameras and projectors make science film making an activity you and the children can easily carry out. If you have fully developed your ability to beg and borrow things from people, you will find that the only expense will be the cost of the Super-8 cartridge and the film processing.

You may be wondering about how film making fits into science class. The medium of film allows you and the children to capture natural phenomena in and out of doors and give them a permanence so they can easily be enjoyed and studied by others. The camera also allows us to study our world without destroying it. Here is a list of things that could easily be preserved in "living color":

1. An active bird's nest (being careful to stay camouflaged so the parent birds do not observe you)
2. Frogs on the edges of a pond
3. The moon on a series of evenings
4. An icicle "growing" on a warm day
5. A spider moving around on its web
6. A comparison of the color of various brooks and streams (pollution??) in your community
7. Farm animals and their young
8. Airplanes taking off and landing at a local airport
9. Finding out whether a heavy and light person on a swing go the same distance from a fixed starting point
10. Butterflies and bees (Be careful!) negotiating for landing space on a variety of flowers

If your classroom is going to emphasize children actively pursuing science, then you are going to have to make sure that you do not dominate the film making experience. If you want perfection, do the films yourself; if you want student learning, have them do the filming.

Free Film Festival

Many government agencies and private companies prepare motion picture films they are willing to send to teachers absolutely free. You have to return them of course after you've finished using them. You need to be wary of one important factor in using films that are free: almost all of them have a commercial. The commercial is the hidden message the agency or company that prepared the film for production wishes to put across. Usually the purpose of the message is to develop within the viewer good feelings about the sponsoring company or agency.

The films are usually well done, so the problems imposed by using free films may be overcome by simply explaining to the children the hidden commercial and letting them know that the people who made the film are also trying to sell an idea of some sort.

One way to select films is to let the children study the film catalogs and vote on those they would find most interesting. They'll be happy to be involved in the process, particularly on the day you announce that *their* film has come.

Companies and agencies that have available free films for use by schools are listed in Figure 10. All you have to do is write a letter on school stationery requesting their latest film catalog. They will be very happy to send you their catalog and film request forms.

Figure 10.
Film Libraries that can supply you with free or inexpensive rental films.

Cooperative Extension Film Library
  Auburn University
  Auburn, Alabama 36830

Division of Libraries
  Pouch G
  Juneau, Alaska 99801

Bureau of Audiovisual Services
  University of Arizona
  Tucson, Arizona 85721

Cooperative Extension Film Library
  University of Arkansas
  P.O. Box 391
  Little Rock, Arkansas 72003

U.C. Agricultural Extension
Visual Aids
  1422 South 10th Street
  Richmond, California 94804

Film Library
Office of Educational Media
  Colorado State University
  Fort Collins, Colorado 80521

Audiovisual Center
  University of Connecticut
  Storrs, Connecticut 06268

Cooperative Extension Film Library
  University of Delaware
  Agricultural Hall
  Newark, Delaware 19711

Motion Picture Service
Florida Cooperative Extension
  Service
  Editorial Department
  University of Florida
  Gainesville, Florida 32601

Film Library
Cooperative Extension Service
  University of Georgia
  Athens, Georgia 30601

Film Library
Cooperative Extension Service
  College of Tropical Agriculture
  University of Hawaii
  2500 Dole Street, Room 108
  Honolulu, Hawaii 96822

Audio Visual Center
  University of Idaho
  Moscow, Idaho 83843

Visual Aids Service
  University of Illinois
  Division of University Extension
  1325 South Oak
  Champaign, Illinois 61820

Audio Visual Center
  Purdue University
  Stewart Center
  West Lafayette, Indiana 47907

Media Resources Center
  Iowa State University
  Pearson Hall
  Ames, Iowa 50010

Cooperative Extension Service
Film Library
  Kansas State University
  Umberger Hall
  Manhattan, Kansas 66502

Audio Visual Services
  University of Kentucky
  Scott Street Building
  Lexington, Kentucky 40506

Cooperative Extension Service
Film Library
  Louisiana State University
  Knapp Hall
  University Station
  Baton Rouge, Louisiana 70803

Instructional Systems
  University of Maine
  Orono, Maine 04473

Audiovisual Service
  University of Maryland
  Room 1, Annapolis Hall
  College Park, Maryland 20742

Krasker Film Library
  School of Education
  Boston University
  765 Commonwealth Avenue
  Boston, Massachusetts 02215

Instructional Media Center
  Michigan State University
  East Lansing, Michigan 48823

Agricultural Extension Service
Film Library
  University of Minnesota
  St. Paul, Minnesota 55101

Cooperative Extension Service
Film Library
  Mississippi State University
  Mississippi State, Mississippi 39762

Audio Visual & Community Services
  University of Missouri
  203 Whitten Hall
  Columbia, Missouri 65201

Campus Film Library for
  Cooperative Extension Service
  Montana State University
  Bozeman, Montana 59715

University of Nebraska
Instructional Media Center
  901 North 17th
  Room 421
  Lincoln, Nebraska 68508

Audio Visual Center
  University of Nevada
  Reno, Nevada 89507

Audio Visual Center
  University of New Hampshire
  Hewitt Hall
  Durham, New Hampshire 03824

Communications Center
  College of Agriculture and
    Environmental Science
  Rutgers University
  New Brunswick, New Jersey 08903

Cooperative Extension Service
Film Library
  New Mexico State University
  Drawer 3AI
  Las Cruces, New Mexico 88003

Cornell University Film Library
  31 Roberts Hall
  Ithaca, New York 14850

Department of Agricultural
  Information
  North Carolina State University
  P.O. Box 5037
  Raleigh, North Carolina 27607

Cooperative Extension Service
Film Library
  North Dakota State University
  State University Station
  Fargo, North Dakota 58102

Extension Service Film Library
  Ohio State University
  2120 Fyffe Road
  Columbus, Ohio 43210

Audiovisual Center
  Oklahoma State University
  Stillwater, Oklahoma 74074

Audiovisual Instruction
  DCE Building
  P.O. Box 1491
  Portland, Oregon 97207

Agricultural Extension Service
  University of Puerto Rico
  Mayaguez Campus
  Rio Piedras, Puerto Rico 00928

Audiovisual Center
  University of Rhode Island
  Kingston, Rhode Island 02881

Agricultural Communications Dept.
  Clemson University Extension
    Service
  Room 92
  Plant & Animal Science Bldg.
  Clemson, South Carolina 29631

Cooperative Extension Service
Film Library
  South Dakota State University
  Brookings, South Dakota 57006

Teaching Materials Center
Division of Continuing Education
  University of Tennessee
  Knoxville, Tennessee 37916

Agricultural Communications
  Texas A & M University
  Room 201, Services Building
  College Station, Texas 77843

Audio Visual Services
  Utah State University
  Logan, Utah 84321

The Audio Visual Center
  University of Vermont
  Ira Allen Chapel
  Burlington, Vermont 05401

Media Services
  Virginia Polytechnic Institute
  Patton Hall
  Blackburg, Virginia 24061

Audio Visual Center
  Washington State University
  Pullman, Washington 99163

Cooperative Extension Service
  West Virginia University
  215 Coliseum
  Morgantown, West Virginia 26506

University of Wisconsin—
  Extension
Bureau of Audio Visual Instruction
  P.O. Box 2093
  Madison, Wisconsin 53701

Audio Visual Services
  The University of Wyoming
  Laramie, Wyoming 82070

Workshops

Center Stage

Through your great abilities as a scrounger of science materials and a very moderate budget ($80) from your principal, you finally have enough materials to operate a sixth grade science program that provides a humanistic learning environment. One of the special pieces of equipment you have is a $40.00 cassette audio tape recorder. During the year your students have used it for such activities as interviewing a doctor in the emergency ward of a local hospital and taping overhead noise in a housing development near the airport, and tomorrow two of the students are going to an interview about water pollution with a local official from the Board of Health. The children have really enjoyed getting out to interview people using the tape recorder. During lunch period you leave the building to fill out a Blue Cross form at the Superintendent's office. You leave your classroom door unlocked so that the children returning from lunch early can come in either to work or to chat. When you return to the room, you discover that the tape recorder is missing. You ask the class about the missing recorder and no one knows anything about it.

1. How do you feel?
2. Do you think any of your students took it?
3. What will the students planning the interview do?
4. Are you going to tell your principal?
5. Will you start locking your room?

Mixing and Matching

Make an illustration showing how you would arrange materials in a room for science teaching. Assume you have these materials: living things (plants and animals), rocks, minerals, fossils, chemicals, hardware, tools, books, paper, desks, chairs, interest centers, odds and ends.

Making a Minibook

Collect several elementary science texts and make collections of minibooks on several topics such as:

1. Animals
2. Sex education
3. Stars and planets
4. Prehistoric monsters

Scrounging

Prepare a scrounge list of things you need in your classroom (have your kids help you). Prepare a letter and then send it and the list to a variety of possible sources for the materials you need.

Observing

Sit in a classroom and look around. What are some things you can do right now to make this room a richer learning environment? Write them down.

Visiting

Visit at least two or three classrooms. How do they acquire and utilize learning sources?

Rearranging

This month let your students decide how the classroom will be organized; let them move things around.

If you don't yet have a classroom, plan one. Make a drawing showing how you would arrange the tables, chairs, resources, etc.

Reference Shelf

ACQUIRING AND USING RESOURCES

Conversation

Joe: This problem of getting resources for science teaching can really be frustrating.

Jack: I think I know what you mean. We spend so much money constructing fancy schools, putting in carpets, special lighting, soundproofed ceilings, and pastel rugs that there is really little if anything left over for science resource materials.

Joe: Even in districts where they're not putting up new schools, inflation and trying to get the rising property tax under control puts a tremendous pressure on school people to keep budgets low.

Jack: So the burden falls on the teacher: put together a super program that is good for the children and keeps the parents pleased, but don't spend money.

Joe: That is the reality and that's why I'm happy that we've tried to share some ways of getting materials that are inexpensive.

Jack: Well, we said that you've got to beg, borrow, scrounge, and just get used to the idea that there will be little money available for materials.

Joe: Is it ever going to get easier? I remember some of the good old days when there were a lot of federal funds around. The first days of school were like Christmas or Hanukkah—opening box after box of science materials.

Jack: Joe, those days are gone forever.

Joe: Yes, but we can dream.

Pastel and India ink.

The child is first.[1]

4
Encouraging Student Progress

The child is an adult in progress, a small human being who is reaching out to the world around him and is in turn touched by it. As he reaches out, he learns. Every second that he exists, he is a transient moving from basic intellectual and emotional levels to more sophisticated ones. Through our efforts in the science classroom we try our best to encourage this process with respect to understanding the natural world. The degree to which we are successful will depend upon:

1. How well we understand how children learn.
2. Our attention to student motivation as a requisite for learning.
3. The techniques we use to help students express their learning.
4. The variety of assessment strategies that are employed.

We hope that this chapter will help you with these four essential steps in encouraging student progress.

How Do Children Learn?

If you are a true student of education who has been formally prepared through courses in psychology, human development, and various other foundations of education experiences, you probably have developed a firm answer to the question posed above. If you don't have an answer yet, we would like to provide you with one, an answer that will be good for all of your days as you work with children, teenagers, adults, lovers, friends, enemies, mothers-in-law, fathers-in-law, trapeze artists, airline pilots, barbers, lawyers, Indian chiefs, riverboat captains, pool shooters. How do humans learn? The answer is:

You pays your money and you takes your choice!

There are as many points of view about the nature of learning as there are experts in the field. Essentially most learning theorists suggest one of two basic positions:

1. The child learns from the inside out.
2. The child learns from the outside in.

Both these points of view will be considered, but one thing remains certain:

Inside-Out Learning

Learning theorists who believe that we learn from the inside out are formally referred to as "gestaltists." As you may know from your previous work, they believe that each human has a unique perception of reality. This perception depends on the way the person organizes the various stimuli coming into his sensory apparatus into a coherent "whole" or gestalt. As you read these words the messages you are getting become part of your perception of reality. The implications of this view are many so we will have to focus on just a few that will be helpful to us as teachers.

If children do indeed have an organized perception of reality, then the process of learning really consists of a change in perception. The child in school who is doing science experiences is a receiver of stimuli: smells, tastes, measurements, or any other messages that enter his being. The key processes of discovering, valuing, and exploring provide the child with stimuli that result in changed perceptions.

A young child might perceive an explanation for thunder that involves "the sound made when clouds bump." As he grows and becomes a keener observer, he "discovers" that he usually hears thunder after he sees lightning. He may place a high "value" on his discovery because finding out that thunder and lightning are related really is kind of interesting. He may even place a sufficiently high value on the discovery to "explore" further. He may ask you more about it, he may want to call a weatherman, he may want to look it up in a reference book. The progression of his perceptions from one level of sophistication to a higher level would be viewed as learning.

Outside-In Learning

Learning can also be explained from the outside in. In this view the child is at the center of an environment, and changes in that learning environment will result in changes in the behavior of the child. (Behavior here is a general term that includes language and writing, as well as other observable phenomenon.) The child is a responder who behaves in ways that assist him in attaining reinforcers. Common reinforcers that exist in the classroom setting include such things as verbal praise from the teacher or peers, nonverbal experiences of approval, smiles, pats on the back.

The outside-in view suggests that learning occurs when the child changes his behavior as a result of the consequences of the behavior. Positive consequences, such as praise and peer approval, increase behaviors; and negative consequences, such as unfounded criticism or ridicule, will decrease his behavior.

The implications of this view for successful teaching and learning are clear: if we wish to help a child progress and grow, we must be sure that we provide an environment that is positive, one that supports and encourages. We must be sure that we provide ample opportunity for the child to express himself so that there is ample opportunity for the natural positive consequences of behavior to affect the child.

Inside-Out and Outside-In

As teachers with 20 or 30 children in our charge within a 20' by 30' room for 7 hours a day for 180 or so days a year, we really have little time to spend theorizing about the nature of learning. We are idealists but we are also practical.

Our life experience tells us that one philosophical view will never provide us with the infinite wisdom we will need to help all our children progress. We will, in the final analysis, let philosophers and psychologists debate the issues surrounding the inside-out (gestaltist) or outside-in (behavioral) points of view. What we need are some principles of learning that bridge the extreme positions and give us direction so that we can truly foster student progress. In the next sections we will consider various learning principles that can help you provide science classroom environments that are both humanistic and effective.

Motivation

"They just aren't motivated."
"She is just so apathetic."
"He doesn't want to do anything."
    Here is a multiple guess question for you. Pick the person(s) most likely to have said all of the above.

    a. Mother
    b. Father
    c. Lion Tamer
    d. Science Teacher
    e. a & b
    f. a & c
    g. b & c
    h. b & d
    i. a & d
    j. c & d
    k. All of the above

Did you pick "k"? Good for you. Lots of people are concerned with motivation; however, we will focus on the science teacher.
    Motivation is the general tendency or disposition toward learning. Motivation is a prerequisite for learning. Just as peanut butter goes with jelly, children with sticky lollipops, and the Lone Ranger with Tonto, motivation must exist for learning to happen.
    One method schools have used to motivate students is *fear*. In the old days a hickory switch took care of the question of motivation. Fortunately, all of the hickory switches have been broken or eaten by termites; but unfortunately, the hickory switch mentality lingers on. In some contemporary schools, children still are motivated through fear, fear of failure, fear of ridicule, and a pervasive fear of all the things that teachers can do to them if they don't learn.
    As practical people interested in encouraging student progress, we need to develop ways to help motivate children. We must find answers to the question: *What things can I do that will get students "hooked" on learning science?* The things that we can do depend upon our ability to organize a classroom environment that reaches out and touches the minds and lives of children. Clearly, a child who learns science in a barren, lifeless classroom is probably going to be less motivated than the child in a classroom that has four gerbils, two guinea pigs, an aquarium, teaching bulletin boards, manipulative science activities, and a lively, enthusiastic teacher. To hook kids on science we must think of the learning environment as our first order of business.

Motivation and the Physical Layout of the Classroom

If you are teaching in a new education facility, you probably have considerable flexibility as far as space utilization goes. Even in older schools there exists more freedom to organize a productive learning environment than most people realize. There is only one problem that is just about insurmountable in terms of the physical layout of the classroom: desks that are bolted down to the floor in rows. If you find yourself in this situation, we respectfully suggest that you get a hacksaw.

A motivating classroom arrangement is one designed to give the child many opportunities to interact with the "things" and "stuff" of science and to interact with other children. You will need to experiment with a variety of seating arrangements until you find those arrangements that provide the largest number of opportunities for motivation.

Some teachers like to arrange the desks and chairs in a wide "U" (Figure 11) which leaves an abundance of floor space for science activities. Other teachers have the children push groups of four desks together to form science stations in different parts of the room (Figure 12). If you are fortunate enough to have tables and chairs instead of desks, you might try one of the arrangements shown in Figure 13.

You may find that children enjoy developing their own floor plans for the room. They will come up with arrangements that are better than any shown in Figures 11, 12, and 13. You might even have the "floor-plan-a-month club" whereby groups of children's ideas are tried on a rotating basis.

Figure 11. Desks arranged in a wide "U."

Figure 12. Groups of four desks pushed together to form science stations.

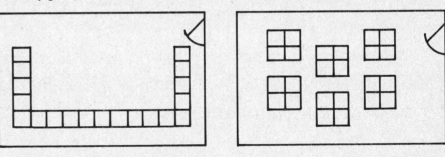

Figure 13. Two room arrangements using tables and chairs.

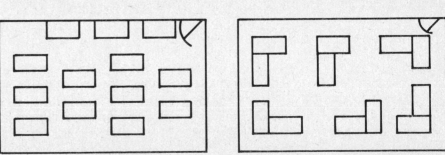

Motivation Through the Display of Science Materials

Many teachers of science try to have a classroom environment that reaches out to students through the display of science materials and equipment (instead of leaving it in the cupboards). Giving kids an opportunity to become motivated as a result of interacting with materials is an excellent idea. Sometimes the approaches we try can be improved upon. Figures 14, 15, 16, and 17 may provide you with some ways of raising the motivation levels of children by simply making a few changes in the way that you "show off" science materials.

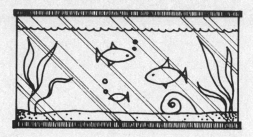

Figure 14. An aquarium alone is nice, but an aquarium with a question card is nicer.

Figure 15. A poster with magazine pictures is nice, but a poster with snapshots of various children in the class is nicer.

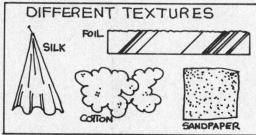

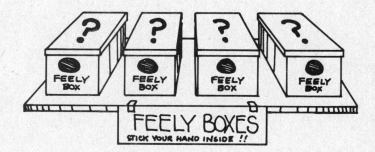

Figure 16. A display of different textures is nice, but a collection of "feely boxes" (children stick their hands in without knowing what's inside) is nicer.

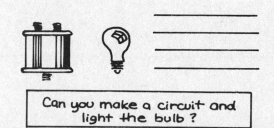

Figure 17. A working circuit that can be switched on or off is nice, but leaving the materials out so that children can make their own circuits is nicer.

Motivation Through Good Teacher Talk

Great actors, great preachers, and great teachers can inspire their listeners. They can lift people from a state of low motivation to a higher level of enthusiasm. Actors, preachers, and teachers can also take a lively, enthusiastic audience and slowly but surely put them to sleep. Boredom is seldom considered a terminal illness, but then think for a moment: have you ever heard a child say

Yes, children can be motivated through our verbal behavior and become ready to learn. Their bright eyes and open ears are eager receptors of that which we wish to share with them. Our verbal behavior can build upon this receptiveness and can help lead them to think things they have never thought and to do things that help them open up and explore the natural world that surrounds them. Sometimes our verbal behavior or "teacher talk" can work against our goal of helping to motivate children. We need to watch out for a few things as we talk and work with children. We need to be sensitive to:

1. Too much teacher talk
2. Too many recall questions
3. Too many statements that are overcritical.

Too Much Teacher Talk

We need to be very sure we don't fall victim to the occupational hazard that can affect all of us (including the authors), a "runaway mouth." It somehow sneaks up on us when we find ourselves in a classroom with a captive audience: we talk and we talk and we talk and we talk. We are seldom content to say, "I have the materials here for an experiment you might want to try." Rather, we begin an elaborate prelaboratory explanation in which we explain what they are supposed to do, what they are supposed to work with, what they are supposed to discover, and what they are supposed to do after they discover what they are supposed to discover.

It is very much like giving a friend a beautifully wrapped gift, telling our friend not only what is inside the package but also how to unwrap it. Too much talk can take the fun out of a lot of things—a good dinner, a beautiful sunset, and a fine classroom.

Too Many Recall Questions

Asking good questions of children can also motivate them to learn. We ask lots of questions in the classroom, but many of them require only the recall of facts. Facts are important. They serve as the building block of science and provide the foundation on which good science and science learning rest. We can, however, overdo questions that are only at the recall level. A useful way of improving our questioning is by thinking about the diagram in Figure 18.

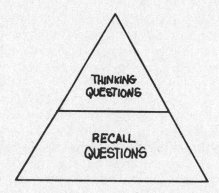

Figure 18.   Questions pyramid.

Thinking questions are more provocative than recall questions. They require the student to put his brain into gear. They ask him to "make a guess," compare, judge, put together, or apply. Here are some examples of thinking questions and recall questions:

Thinking	Recall
What would happen if we gave the fish too much food?	How many guppies are there in the fish tank?
How could you use the seesaw (teeter-totter) with someone much lighter than you?	What do magnets do?
If snow were black, would eskimos have problems with their igloos?	Do birds fly south for the winter?
How could we find out what weathermen do?	Is green a primary color?

A good mix of thinking and recall questions can be a great help to you as you try to help motivate children. Emphasizing just recall questions will simply teach the children that science is a collection of facts when it is, in reality, far more than that.

Too Many Overcritical Comments

"Teacher talk" that is overly critical can also work against your other efforts at motivation. Try your best to accentuate the positive in all your dealings with children throughout the day. Try to find something good to say to every child every day. Children grow and progress much better in a positive, supportive environment than in a negative, overly critical one. Where ever possible we should emphasize the "warm fuzzies" and cut down on "cold pricklies."

Learning

The motivated child is a child who is ready to commence learning. Learning is an interactive process in which the learner works on his environment through a science experience and in return becomes changed by the process. The motivated learner is eager to act and be acted upon. This interaction of the learner with the materials of science and the consequent change in the child's attitudes, knowledge, values, concepts, and manipulative skills is an ongoing process that should have no end. Each interaction elicits from the child new questions and new interests which in turn lead him further down the road to understanding his environment.

In this book we view learning as involving the children in three basic processes: discovering, valuing, and exploring. As the child moves through these processes, his understanding of science becomes deeper and more comprehensive. One can visualize this by seeing the experience of learning as an ever widening spiral (Figure 19).

The child *discovers* a phenomenon, ascribes *value* to that which he discovered, and then *explores* further. To ensure the continuity of experiences necessary, our challenge as teachers is to find and utilize a variety of techniques for encouraging a child's progress. The remainder of this chapter consists of suggestions that will help you assist children as they learn science.

To help a boy or girl learn science in our classrooms, we need to provide the type of science instruction that fosters an active student role. It is true that

Figure 19.   Learning spiral.

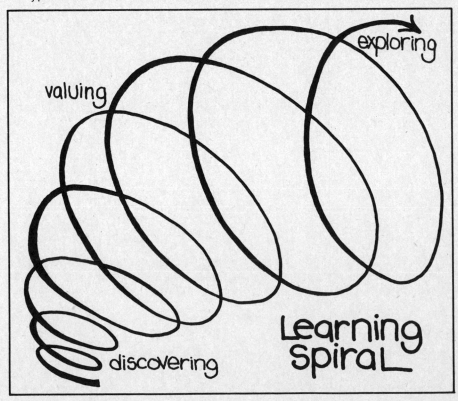

there must be some student time available for reflection and other, more passive endeavors, but the science experience taken as a whole must be one that has the child as an ''overt,'' expressive young human being. Student activity is necessary because:

1.   A child who does something is more likely to retain and use the knowledge, concepts, or attitudes he acquires.
2.   A child who is doing something can be observed and encouraged.

Our common sense affirms the validity of the first principle, but the second principle is one that we must really focus upon if we are to find practical ways to put it into practice. The central question is: *What can we do as teachers to provide opportunities for the child to express what his learnings are?*

Science Logs—Expression in Writing

The science log provides for children a way to express in writing that which they are learning. By committing something to paper, the child gives us something to respond to. Our response to his work can provide the encouragement for continued progress.

The log is the child's medium for expressing through words and/or pictures the results of his science experiences, as well as his feelings about the experience. You may need to help children who have never kept a log get used to working with one. Experienced teachers know that a spiral notebook will probably serve as a better log than a loose-leaf one because of the ease with which loose-leaf sheets seem to leave children's notebooks. The spiral type may even retain its pages for the year (well, would you believe a month?).

Have the children decorate the outside front and back covers of their science logs for quick identification. The critical test of the adequacy of the decoration is whether or not the log can be seen and identified at a distance of ten feet or when covered by five mittens, two scarves, and a catcher's mitt on the floor of the cloak room. The range of artistic ability displayed on log covers will surprise you. You may have one that looks like one of those pictured in Figure 20.

Figure 20.   Science log covers.

Figure 20. Science log cover.

Each time your students begin a science activity, you will need to remind them to take their logs out. Expect the children to date the page or portion of a page they are on and write the name of the activity or general topic under investigation. For young children you may want to write this information on the blackboard so that they can organize their log accordingly.

As the children participate in the science experience, encourage them to write or draw the interesting things they find out and to tell how they felt about them. You can expect long entries that look like those in Figure 21.

Figure 21. Science log entries.

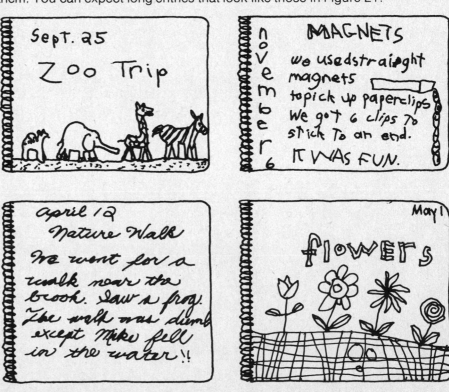

Children's writing or drawing provides you with information concerning what they are learning and how they are feeling. Your use of this information as an opportunity to encourage them is essential. The science log can be a focus for a science conference after an experiment or group of experiments. You may wish to write something in each child's log to further encourage his or her diligent work.

Child Talk—Expression in Speech

Earlier in this chapter we discussed the role of good teacher talk in motivating children. Now we must consider child talk as a factor in their learning. Most children in your classroom will probably be quite capable of expressing themselves orally. When they see you in the morning, they may bubble over with enthusiasm about whatever has happened to them since you saw them last. They will talk and talk and talk. The talk will range across many subjects. You will hear such things as "My fish died," "I saw a rainbow," or "Baby brother ate a grasshopper."

Talk is the medium through which children articulate the results of their interactions with and observations of the world around them. Careful listening to child talk provides us with a method of sensing what the child senses and determining his needs. We must, however, listen very carefully.

That which the child shares with us orally is a precious thing, for it provides for us a stream of information that lets us know what is going on in his mind. We earlier defined learning as a process of change, and we must view changes in verbal behavior as indicators of the strengths and direction of change.

As significant figures in the child's environment, we are in a position to do much to affect positively (or negatively) the scope and speed of his learning. We really have two things to do when a child expresses himself orally:

1. We must listen.
2. We must respond.

Good teachers are good listeners. As good listeners we need to remember to orient ourselves so that the child knows he is getting our full attention. If you work with young children, don't feel it is improper to kneel or squat so that you are in a position to fully attend. Why give a child a stiff neck from having to hold his head at a forty-five-degree angle just to speak? Sometimes we simply forget how tall we are in comparison to young children. Perhaps Figure 22 expresses this point more clearly.

Figure 22. We need to give children our full attention when they are speaking to us.

Children are sensitive to the manner in which we attend to their talk. Even the skilled teacher may find that the demands for attention to any one individual's verbal behavior are difficult to meet because of the number of children in the classroom and that it is impossible to grant large quantities of undivided attention. Both young and older children will no doubt appreciate one or two minutes of undivided attention as compared to four or five minutes of teacher attention shared with others. A child's talk is such an important reflection of his learning that a sacrifice in quantity may have to be made so that we can provide a few moments of concentrated attention for each child in the room.

Our response to that which the child says is of the utmost importance. The child needs to be encouraged following oral verbal behavior. Whenever possible we must follow his or her statements with praise and positive nonverbal behavior as well. One, of course, must consider the problem of what to say or do when the child tells us something that is incorrect or expresses a misconception.

When a child tells us something that signifies the ''learning'' of something that is incorrect, we must at the minimum say something to show that we support and encourage his expressing himself, if not the content of his expression (Figure 23).

Figure 23.   We must support the act of expressing if not the content of the expression.

Assessment

Assessment is a scary word. Some may feel that even talking about it labels one as being opposed to the provision of learning environments that are intended to be humanistic in nature. Assessment is scary if we think of it only in terms of formal testing, comparing children one to another, or punishing children for lack of success. How many "war stories" can you tell about teachers you had who used assessment for the wrong reasons? Yes, assessment is too often used by teachers as a weapon instead of a productive, positive way of diagnosing the effectiveness of our procedures for motivating and teaching.

Assessment does not have to be a negative experience if we take the time to develop a few techniques that bring out its positive, productive aspects. We need to focus on assessing a child's growth in science so that the information gathered gives us the information we need to change our teaching-learning procedures. Let's consider then various techniques for acquiring feedback that will help us plan superb classroom environments for children:

1. Tests, quizzes (some cautions)
2. Observations
3. Science conferences

Tests, Quizzes . . . Some Cautions

As we achieve success in providing a more flexible science program for children, the usefulness of traditional testing procedures becomes very limited. In the conventional classroom with every child studying the same thing at the same time, it was easy for children, teachers, administrators, and parents to accept results from tests and quizzes as an important assessment of student progress. In truth, however, teacher-made tests lack believability. Even a quick perusal of the field of tests and measurements tells us that meaningful information can result only if tests meet the twin criteria of validity and reliability. We really don't have space to expand on these terms here, but suffice it to say that most of us as teachers lack the time to get involved in developing good tests. This lack of time, coupled with the increased choice making available to children in humanistically oriented classrooms and the potentially destructive effects of a strict testing program on a child's attitudes, make the utilization of tests for children in the science area a questionable pursuit in the early elementary grades.

In grades 5, 6, and 7, tests may be appropriate if they are developed jointly by the youngsters and you, and if they are viewed as a positive assessment procedure. If the grade aspect of tests and quizzes is played down and their diagnostic potential is emphasized, they can play a minor role in student assessment. Other techniques, such as direct observation and science conferences, should be the major assessment strategies used.

Observation

When your boys and girls are actively engaged in science activities, they are providing a wealth of information for you about the quality of the motivation and learning environment that have been provided. Most of this information is lost because you can't possibly see and hear everything that is going on. Nonetheless, you can still manage to gather enough information to make some judgment about how well things are going. Let your eyes and ears take in as much as they can.

Figure 24. Classroom observation chart.

OBSERVATIONS*	Monday	Tuesday	Wednesday	Thursday	Friday
1. Children working in groups.					
2. Number of students working independently.					
3. Number of students seemingly bored.					
4.					
5.					
6.					
7.					
8.					
9.					
10.					

ANECDOTAL RECORD:

Week of _____

Period _____

* You may wish to list your own observations. We suggest you refer to the SCAN card and science classroom checklist in Chapter 5 for additional observations.

Classroom Observation

You cannot observe very well if you are seated at your desk with your mind engaged in contemplating a crossword puzzle or the latest copy of *Europe on 26¢ a Day* or *How I Found Truth While Climbing the Himalayas.* The teacher has to be an active observer. You need to be on your feet roaming around the classroom, seeing (not looking), listening (not hearing). You must think of yourself as an information gatherer getting through your senses answers to specific questions, such as:

1. How many of the children are focused on the activites?
2. Are there any children who are just recording the information done by the experimenter? (Secretarial chores should be shared.)
3. How many happy faces are there?
4. Are any of the children doing nonscience work (such as assignments from other subject areas)? If so, why?
5. If the children are supposed to be keeping science logs, how many logs are visible?
6. What are the children talking about?

As you are moving around, talk to the boys and girls, and listen for the "child talk" described earlier. Encourage their work and ask questions. From time to time ask children when you are near them such questions as:

1. How are you doing?
2. Are you learning anything?
3. Do you like this?
4. .Do you need any help?
5. What do you want to work on next?

The key to successful observing is your active involvement in what is going on in the classroom. This, coupled with focusing upon specific questions such as those suggested above, will provide a wealth of useful information. You might wish to keep track of your observation by preparing a simple record-keeping chart, such as that shown in Figure 24, which you can use once or twice during each science period. Simply list the observations you wish to make vertically and the days of the week horizontally. Try to discipline yourself to write down a comment under the anecdotal record for each day so that you have general reactions to the day.

This weekly form can provide you with a great deal of information. Be sure to record the observational data as well as any anecdotal record you may wish to keep on a daily basis. Keep the forms in a folder or notebook so that you can have a permanent record of how things are going.

Individual Observation

Just as we need to assess the overall impact of our science learning environment, we must also focus upon the individual student. The observation techniques expressed above fit nicely into an assessment strategy by which you can gauge an individual's progress. Simply devise a record-keeping form you can use to record your observation in a quick, efficient fashion. List the children's names vertically and the days of the week horizontally as in Figure 25.

You may find that keeping track of every child every day with an anecdotal record form of this type early in the year will use up too much of your time. This is especially true if you are involved in other tasks that must be carried out for science class to go well. You might wish to prepare a master form as shown

Figure 25. Individual observation chart.

INDIVIDUAL OBSERVATION CHART

WEEK OF ___10-22___

STUDENT	MONDAY	TUESDAY	WEDNESDAY	THURSDAY	FRIDAY
Bobbye H.	Working well	doing drawings	in argument with Steve H.	working away from group	back in group
Louie N.	Planted seeds	Putting newly planted seeds in closet	Checking seeds in closet	Put some seeds on window-sill	Bored... nothing happening with seeds
Bill C.	Very sleepy today	absent	absent	very sleepy	Absent (call parents p.m.)
Steve H.	Leading group	very bossy today	in argument with Bobbye H.	working away from group	absent
Kitty A.	Doing science drawings	more drawings	more drawings	started seed-growing experiment	reading minibook on seeds

ADDITIONAL COMMENTS:

above but use it to record observations for just those students who may need extra time and attention. Your day-by-day record of their work can provide an excellent basis on which you and the children can plan their future work.

As the year goes on you will find that you can expand the use of an anecdotal record-keeping system to include all the children. You may even be able to have the children keep their own daily record as part of their log. You can provide them with a blank form to be pasted in their log so they can easily record their own daily progress. This utilization of an anecdotal record as a self-assessment technique can help children learn that the ultimate responsibility for self-growth rests with them.

Science Conferences

What is so rare as a few minutes to sit and chat with a child about how he or she is doing, feeling, and thinking, a chance to engage the child in a one-to-one interaction that lets him know that you care about him, his growth, and his feelings? The importance of individual conference time cannot be over-emphasized. Think of yourself as part of a group of twenty or twenty-five children. You each get along, yet each of you wonders if anyone really cares. You begin to think: "Teacher says that we are doing well, but she never has a chance to talk to me about what *I'm* doing." We must be sensitive to the development of this attitude of despair that can easily transform a happy, productive child into an apathetic, resentful one. One important teaching technique you can use to keep in touch with the individual child is the scheduled science conference.

Because of the very busy schedule we teachers have, science conferences with children must be designed around both the teacher's and the child's schedule. One approach you might try is to review your daily schedule and perhaps find two or three five-minute blocks that would lend themselves to conference time. There need to be times when the remainder of the class is working on matters that do not require your direct attention. Perhaps you could identify at least one time block during science period, one during free reading time, and one during arts and crafts time.

For efficient utilization of these brief time blocks, you will need to post a schedule and sign-up sheet in a highly visible place in your room and request that the children schedule themselves. The number of conferences you can have each month with each child obviously depends on the number of children you have in science and the number of time blocks you can find. It is absolutely essential that the children understand that signing up for a science conference is their responsibility. Of course, they should also understand that you are ready, willing, and eager to talk with them at any time and that the conference system is an additional opportunity for them to get your individual attention.

The content of the conference should depend on the child's individual needs. Early in the year you might wish to focus the science conference on such questions as:

1. How are you doing?
2. What are you working on?
3. What can I help you with?
4. How are you feeling about science?
5. How is your science log coming along?

As the year progresses the children should take more initiative for the content of the science conference and your role should become increasingly less direct. Your goal should be to help the children become comfortable enough with you and science so that you need only ask, "What would you like to chat about?"

Workshops

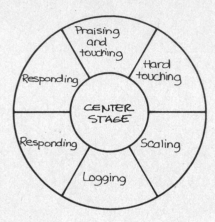

Center Stage

At a meeting with faculty in your school, you announce that you are going to let your students decide the science topics they will study for the second half of the year. A younger teacher jumps up and tells you with much emotion that you are throwing away the professional judgment you were trained to use.

What are your feelings? What are you feeling toward this teacher? How will you defend what you plan to do with your class?

Praising and Touching

Sneak up on someone today and praise him or her. What kind of reaction did you get? Do you ever touch people when you talk to them? Are you comfortable touching people? Do you think praising or touching others has much to do with their learning?

Hard Touching

Have you ever been hit by a parent or teacher? How did you feel about the person who hit you? What was your immediate reaction? How did you feel about it later—when you had a chance to think about what happened?

Scaling

Make a rough scale drawing of how you would arrange the desks or tables for science in the following classrooms:
1. 24 student desks
2. 6 tables (seating 4 students each)
3. 2 tables (4 students each) and 16 student desks

Logging

Write out an explanation that you would give to children about how to keep a science log.

Responding

During a science conference, a child tells you the following:

*"My dog chewed up my science log, someone stole the plants I had
growing, I hate doing science, and besides you always pick on me."*

What would be your very next statement?*

Responding

A child has just told you the following:

"My mother said that if you touch a frog, you will get warts."

What would you say in response?

Reference Shelf

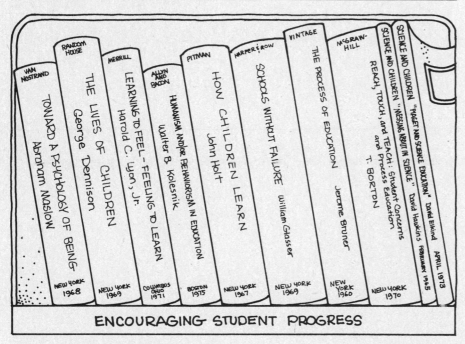

ENCOURAGING STUDENT PROGRESS

Conversation

Jack: Well, we've tried to point out that student progress happens as a result of a sort of three-step process: motivation, learning, and assessment. How well do you think we did?

Joe: I think we at least hit the high points in each of the three areas. We meant it as an overview, and we do have some pretty good examples.

Jack: But there is so much more about motivation, learning, and assessment that we could have included.

Joe: I agree, but we do have to assume that the good folks working with the book come to it with some background or experience in the areas that we mentioned. Anyway, the major point of the chapter is that motivation, learning, and assessment are tied together. I hope everyone will try to keep that in mind. Jack, what if teachers want to learn more about these topics?

Jack: I suppose those that are motivated will look on our Reference Shelf.

Joe: I set you up pretty good for that one.

Jack: Thanks.

Joe: Well just don't be like the teacher on the ecology field trip who took everything for granite.

Jack: Or the bad little boy on the science field trip to the beach who left no tern unstoned.

Joe: This has got to stop.

*Now, now remember you are a teacher, so delete those expletives.

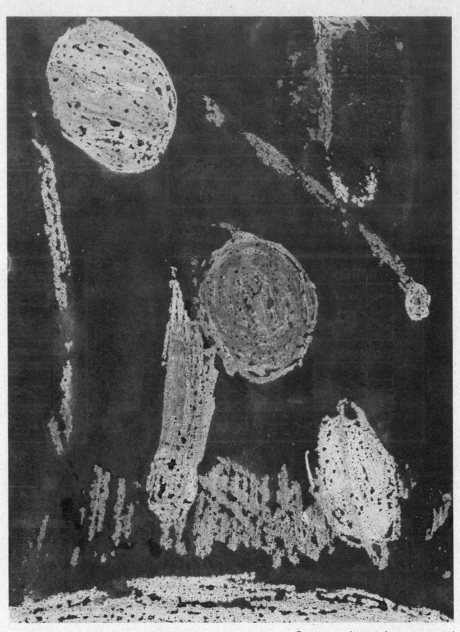

Crayon and paint (crayon resist).

5
Becoming
More
Humanistic

Do you

- sometimes feel frustrated about teaching?
- sometimes experience feelings of loneliness?
- at times wonder if your students are learning anything?
- wonder if other teachers experience the same feelings as you do?
- wish you could discuss your problems of teaching with other teachers?
- sometimes wonder how you are doing as a teacher?

If you've experienced any of these, don't despair. Teachers we've worked with who are attempting to make changes in the manner in which they teach have experienced many of them. (We think we've experienced them all!) We view asking questions such as these and experiencing these feelings as a sign of a healthy, growing person. We will not change and grow unless we face ourselves. We as teachers must probe and try to understand what is happening to us. When we dare to do so, we are becoming more humanistic.

Carl Rogers would view these concerns as an attempt on our part to communicate what we are experiencing. In a sense it is another way of being congruent or real. Rogers makes an important point when he talks about realness.

What do I mean by being real? I could give many examples from many different fields. But one meaning, one learning is that there is basically nothing to be afraid of when I present myself as I am. When I can come forth nondefensively, without armor, just me. When I can accept the fact that I have many deficiencies, many faults, make lots of mistakes, am often ignorant where I should be knowledgeable, often prejudiced when I should be openminded, often have feelings which are not justified by the circumstances, then I can be much more real. And when I can come out

*wearing no armor, making no effort to be different from what I am, I learn
so much more—even from criticism and hostility—and I am so much more
relaxed, and I get so much closer to people. Besides, my willingness to be
vulnerable brings forth so much more real feeling from other people who
are in relationship to me, that it is very rewarding. So I enjoy life much
more when I am not defensive, not hiding behind a facade, just trying to
be and express the real me.*[1]

How often and to what extent are we real when we think about our teaching or
the students we teach?

This chapter is designed essentially to provide you with some help in dealing with the question: *How can I find out how well I am doing?* Dealing with this
question is not easy, and in fact may be painful. Facing ourselves head on as
teachers makes us vulnerable, but at the same time ready for growth and
change. This is at the heart of becoming more humanistic. We see this process
as being intrinsic as opposed to extrinsic, operating on internal standards
rather than external, a clarifying one rather than a fixed one. We do not feel
there are any standards you can evaluate yourself against. When we discussed
student assessment in the last chapter, we emphasized assessing a child's
growth. We asked that you focus on observation and personal communication.
We feel that evaluating your own growth is no different and will make you more
sensitive to the process of growth in general. We are going to encourage you to
use self-evaluation as the primary process in answering questions about your
teaching.

To help you clarify your position with respect to teaching and learning, we
provide in this chapter a collection of self-evaluation inventories and questionnaires, and suggestions on how to use student and peer feedback to assess
your own progress.

At the end of the previous chapters, we have carried on a conversation between ourselves. In this chapter our conversation includes the three additional
voices of teachers who face the realities of day-to-day experiences with children. We hope this will give you an opportunity to touch base with other teachers who are attempting to deal with science in a humanistic setting.

Self-Evaluation

An Inventory

We start with you. The first thing we want you to do is to take a test (Figure 26),
but this test has no right or wrong answers. The questions are about some
ways of operating in a humanistic classroom. Take some time and do this now,
before going on.

After taking the test examine your responses in the following ways:
1. How did you feel when you finished the test?
2. Were you honest?
3. What did you learn about what you are now doing as opposed to
    what you would like to do?
4. Based on your responses about what you are now doing, how
    would you characterize your classroom environment?

You may want to return from time to time during the year and check yourself
against this test.

Figure 26. Open Learning Environment Test.[2]

TAKE THIS TEST

The following statements are about some ways of operating a classroom. Please respond to each statement under 2 categories:

A. What you are NOW doing.
B. What you would LIKE to do.

FOR EACH STATEMENT CIRCLE THE TWO MOST APPROPRIATE DOTS.

	NOW DOING		LIKE TO DO	
	NEVER SELDOM SOMETIMES OFTEN ALWAYS		NEVER SELDOM SOMETIMES OFTEN ALWAYS	

1. Each student can decide for himself whether or not to take part in a particular assignment.
2. Each student is allowed to decide how he will study each topic.
3. Each student determines how much time he spends on a topic.
4. Students are free to group as they want.
5. Every student has free access to all the materials in the classroom.
6. Students determine what is removed from or added to the classroom.
7. Every student is completely free to move in the classroom.
8. Students freely ask for help whenever they need it.
9. Students are graded on a curve.
10. Students have the responsibility to evaluate themselves.
11. Students self-initiate their own activities whenever they choose.
12. Each student has his own personal space in the classroom (drawer, cupboard).
13. Peer-group teaching is a primary activity in the classroom.
14. Many diverse activities simultaneously go on in the classroom.
15. Students are encouraged to report on topics in any way they want.
16. Do you have contacts with the students about their learning?
17. I am confident my students will learn if left to themselves.
18. I leave my students alone in the classroom.
19. I lower student grades when they make mistakes.
20. Do I follow a school outline, a manual, or a text in pursuing course content?
21. Students have fixed places to sit.
22. My class is child-centered rather than subject-centered.
23. Students can leave the classroom whenever they want to.
24. Students are free to bring anything they want into the classroom.
25. Students are encouraged to consult other teachers as resource people.
26. I encourage teachers, parents, administrators to enter class as resource people & observers.
27. The operational rules of the classroom are made by students.
28. Students are free to do nothing.
29. Individual novel solutions are rewarded more than concensus solutions.
30. Students are encouraged to develop personal goals.
31. I talk individually with students about their personal goals and then pursue their development.
32. Students are free to talk to each other at any time.
33. Individuals are encouraged to pursue their own interests.
34. I like to go to school in the morning.
35. Students consider their total community as a primary resource.

Listening to Yourself

Another self-evaluation tool is that of listening to yourself and your students. (If you have access to a video tape machine, you might wish to use it.) In most instances a portable cassette tape recorder will provide you with all that you need. We suggest that you set the recorder in a place in your classroom, turn it on, and proceed with your normal routine. Record several episodes of teaching during the day, week, and year (each 20 to 30 minutes in length).

The first thing you should do is just listen to the tape. Don't impose someone else's classroom behavior instrument on your own teaching behavior. The reason we say this is that they represent someone else's window or vantage point from which to view classroom behavior. Later you may wish to look at your behavior from other vantage points, but now is the time to become freely in contact with how you are in the classroom.

After you have listened to an episode of teaching, you may want to respond to some or all of these questions:

1. Whose voice was the dominant one on the tape?
2. What types of statements did you hear yourself saying: questions, commands, instructions, praise, negative comments, etc.?
3. How many different voices did you hear?
4. What types of questions did you ask, low cognitive level (such as what is the name of this), high cognitive level (such as why do you think that happened), or value questions (such as do you think that is a good idea)?
5. Did the students ask you questions? How many? What kind?
6. Do students appear to be enjoying themselves in your classroom? Are you?
7. Do you initiate activities, or do the students?
8. Are some student conversations not related to school? How do you react to this?
9. How do students respond to your questions?
10. Do the students seem motivated to do science in your classroom?

These questions should serve as stimulators for you to ask further questions. After reviewing your tape, use these questions to
1. Identify some of the patterns of your classroom behavior.
2. Identify changes you wish to make in your behavior.

In the next section we have included two instruments you may wish to use to study your classroom teaching behavior in more detail.

Teacher Logs

Finally, we suggest that you try an additional technique. At the end of each day for about two weeks, write a brief account of what happened during that day and enter it in your teacher log. Try to recall significant events, how you felt about the day, and whether you thought things went well. Include problems you see developing and solutions you think may solve them.

This method can provide you a rich source of information with respect to your view of teaching, problems you perceive, and possible solutions to them.

Student Feedback

"She's a fantastic teacher!"

"Ugh! I don't want to go to his class today!"

"I wish she wouldn't talk so much."

"I like the way he treats us, you know, he sort of cares for us."

"We never do anything here."

"I think we'd learn a lot by going outside!"

The degree to which we use feedback from our students to help us improve how we teach is a measure of our openness. Openness is not easy. It's not easy to open ourselves to others. But for so long we've paid lip service to what our students say. If we are really honest with ourselves, we'll begin to realize that our students see us in action each time we teach. Who better can tell us how we are doing!

Throughout this book we've encouraged you to risk, to grow, to love. We think student feedback is important. It's risky. But students, particularly if you are moving toward a more humanistic classroom, will be honest with you. We think if you try it, you'll like it.

In this section, we are providing several alternatives in terms of how to obtain student feedback. We recommend that you select one that you feel comfortable with, and use it. Later you may wish to use several, combine some, and develop your own.

The Questionnaire (For Older Kids)

Questionnaires are useful if they are related to you and your classroom. Two different questionnaires (Tables 5 and 6) are included. You could use either, or combine both by selecting individual items to make your own instrument. Investing 10 minutes will provide you with impressions your students have of your teaching and how they feel about your classroom.

Table 5

Student Questionnaire[3]

	Reaction			
Circle the number to indicate whether you: 1. strongly agree 2. agree 3. disagree 4. strongly disagree	Strongly Agree	Agree	Disagree	Strongly Disagree
Statement				
1. I like science.	1	2	3	4
2. My teacher allows us to make suggestions about what we study.	1	2	3	4
3. My teacher seems excited about teaching science.	1	2	3	4
4. I cannot learn unless I am supervised.	1	2	3	4
5. This class is very worthwhile.	1	2	3	4
6. I can do science without knowing much about the subject.	1	2	3	4
7. I can talk with the teacher as a person.	1	2	3	4
8. I should not be allowed to choose who I work with.	1	2	3	4
9. My teacher seems interested in what I have to say.	1	2	3	4
10. I can trust the teacher.	1	2	3	4
11. This class is often boring.	1	2	3	4
12. We seem to have freedom in class.	1	2	3	4
13. My teacher is good at presenting material so I can understand it.	1	2	3	4
14. I like what goes on in this classroom.	1	2	3	4
15. The activities we do are interesting.	1	2	3	4
16. I think we should grade ourselves.	1	2	3	4

Faces[5] (For Young Kids)

You can use the Faces technique at various times, such as at the end of a lesson, a week, and so forth. Distribute a copy of the form (Figure 27) to each student. Select three questions you wish your students to respond to. If it is at the end of a lesson, you may ask questions such as:

a. How did you feel about working with your partner today?

b. How did you feel about learning about magnets today?

c. How do you feel about doing an activity like this again?

Have your students select a face that indicates how they feel about each of your questions.

The questions you ask will depend upon when you use the form, and for what purposes. Reexamine the questions in the questionnaires in Tables 5 and 6 for additional questions. You can also use Faces to get at the process of valuing which constitutes the substance of Chapter 8.

Incidentally, this can be used with older kids, too!

The Letter

At the end of about two or three months of your class, ask each student to write you a letter in which each responds to the following:

I wish to have you write me a letter in which you evaluate my classroom. I would like you to tell me about the positive aspects and the negative

Figure 27. Faces.

Table 6.
Tell Me What You Think [4]

Statement	Strongly Agree	Agree	Disagree	Strongly Disagree
1. I like this class.	1	2	3	4
2. I think most of the kids enjoy this class.	1	2	3	4
3. I would like to do more independent work.	1	2	3	4
4. I think we should do more group work.	1	2	3	4
5. I think we should take more field trips.	1	2	3	4
6. I think your evaluation system is good.	1	2	3	4
7. I think we should do more valuing activities.	1	2	3	4
8. I think the opinions of other kids in this class are important.	1	2	3	4
9. I think we should have class meetings to discuss class problems.	1	2	3	4
10. I think we should be allowed to do more self-evaluation of our work.	1	2	3	4
11. I have warm feelings about this class.	1	2	3	4
12. I feel we learn more than just science in this class.	1	2	3	4
13. I think ample time is provided to do our work.	1	2	3	4
14. I think the class senses how you respect each person's ideas and feelings.	1	2	3	4
15. I think this class is a good place to learn.	1	2	3	4

aspects of the class. I would also like you to make any suggestions for improving or changing the class. I want you also to know that this will have no affect on your grade. Thank you.

You can ask the students to sign the letter or leave it anonymous. This will depend on where you are with your students in terms of the level of communication in your classroom. For some kids, signing the letter is risky. You'll have to use your judgment.

Interviews and Conferences

This method departs from the first three in that now we are suggesting that you sit down face to face with your students and carry on a discussion focused on obtaining feedback about their feelings and attitudes about your class.

The conference we are suggesting here is similar to the student conferences suggested as a process of student evaluation; however, it differs in one very important way. You will now be asking the student, "How am *I* doing?"

This is really moving beyond loving because now you're saying to your students that they are human beings, and that you value what they say. You are also saying to them that you trust them, and that they can trust you. Trust takes time, and you'll have to invest much time and energy to attain it.

The main ground rule to follow in a personal conference or an interview is that the students have the right to refuse to answer a question, and that they can ask you the same questions. You also have the right not to answer their questions.

The questionnaire presented in this chapter could be used as a source of questions if you are interviewing students. Also, you are in a position to follow up on each student response, and this will enable you to ask additional questions. After the conference is over, you should write in your teacher log a summary of the conference including:

· your personal reaction or feelings
· suggestions made by students for change
· problem areas

A variation on the conference is to divide your class into groups of about five and ask them to evelute the class. The statement included with the open-ended letter method could be used as a focal point for discussion. You could ask that each group provide you with a brief report, or you could meet one member from each group for a conference.

Peer Feedback

Many of us close the door when we teach! We psychologically shut our colleagues out from our classroom and from the manner in which we teach. We feel you may find it valuable to discuss observations of your teaching with your colleagues whether you are in a team teaching setting or in a self-contained classroom.

Figure 28. Two classroom scenes.

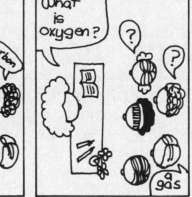

Several methods can be employed to obtain peer feedback; but regardless of the method, the feedback should conform to the following ground rules: ·

1. Feedback should be descriptive. Describe what happened in the classroom, not what you think should have occurred.
2. Feedback should be nonevaluative. If it is descriptive, no evaluation (good or bad) can be interpreted. If something went wrong, the teacher knows it, and you probably won't have to tell him.
3. Feedback consists of suggestions for alternative ways of doing something.
4. Feedback should be simple. At any one time, an individual can change only one or two things. If you hit him with a barrage of points, he will probably become confused and will not be able to change anything. Keep feedback to a few main points, such as the types of questions asked or the behavior of a few students.
5. Feedback belongs to the person you observed. If you kept notes during the playing of the tape or while you visited a class, give them to the person you observed at the end of the conference. You should never write something you do not want to show him or her. [6]

Feedback about Your Classroom Behavior

Imagine the two classroom scenes shown in Figure 28. What kind of feedback would you give each of these teachers? Perhaps the feedback shown in Table 7 is what you might have given.

Table 7.

Peer Feedback		
Feedback	Teacher 1	Teacher 2
Questions asked	3	1
Type of questions	3 closed	1 open
Type of student talk	One-word response	Lengthy response
Number of students talking	4	3
Teacher nonverbal behavior	Standing in front of students	Sitting with students

You might be wondering why we didn't make remarks such as:

·   The students are interested in what they are doing.
·   What a lousy lesson!
·   Boy, these kids are having fun!
·   He really knows what he is doing!

We could have, but we want to make a point. The feedback listed for each teacher is a set of observations. You probably would not doubt that Teacher 1 asked three questions and Teacher 2 asked one. But would you agree with the statement (about Teacher 2) "What a lousy lesson!" Probably not. Such a statement is an opinion, or an inference about teaching. We don't want you to remove opinions, but rather have you start with observations about teaching and then make your evaluation.

We hope to provide you with feedback about the following two aspects of classroom behavior:

1. Questioning
2. Your students' behavior

You'll note that our feedback statements were basically confined to both of these.

Questioning

This is probably one of your more powerful behaviors in terms of tapping your student's curiosity and interest in science. In the Kid Stuff portion of this book, many of the activities give you an opportunity to ask meaningful and interesting questions.

We can think of questions in two ways:

1. Closed—These are questions we ask that have a limited number of acceptable responses (Scene 1).
2. Open—These are questions we ask that have a greater number of acceptable responses (Scene 2).

STOP reading the book now and do one of the following:

1. Tape record one lesson to determine whether you ask predominantly closed or open questions.
2. Reflect on the last lesson you taught or observed. Were the questions closed or open?
3. If you are taking a course, does your teacher ask closed or open questions?

Patricia Blosser has developed an observation scheme for questioning which we have found useful. Table 8 shows the various categories in the system.

Remember the pyramid we used while discussing *good* teacher talk in Chapter 4:

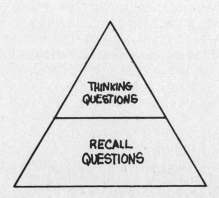

Table 8.

Question Category System for Science[7]

Closed	What the Teacher Says	What the Student Does
1. Memory	What kind of rock is this? Where does the bird fly in winter? Which end of the compass needle points north?	recalls repeats identifies names observes points
2. Convergent Thinking	How would you arrange these objects to form two groups? What do you think will happen if the temperature drops below 0°C tonight?	classifies applies predicts
Open		
3. Divergent Thinking	What do you think the inside of this object looks like without looking in? Do you think there is life like ours elsewhere in the universe?	gives an opinion infers implies creates reorders
4. Evaluative Thinking	Do you think it's a good idea to build nuclear power plants in our city? How are you going to prove your point?	justifies a point designs a hypothesis or conclusion makes a value judgment

We wish to remind you that thinking questions are more provocative than recall, and we should try our best to strike a balance between thinking questions and recall questions.

To help you use the Blosser system, we have included some sample questions ("What the teacher says") and some student behavior ("What the student does") for each category. We have found it helpful to study these first, and when you teach a lesson focus on one or two categories (such as memory and divergent) rather than all four categories.

Questioning is a skill you can learn and improve upon. Having a colleague visit your classroom and record the types of questions you ask is a technique we recommend. There is an art to asking the right type of question at the right time. Perhaps Table 9 will provide you some guidance in developing your own questioning strategies.

Table 9.

Improving Your Questioning Skill*

Purpose	Meaning	Examples
Diagnosis	To detect and analyze the progress of students in learning situations.	Asking a series of memory or convergent questions (sometimes called *probing*).
Clarification	To clear up any learning problems the students may be having, making the materials more understandable.	Asking a series of memory or convergent questions. Asking the student to rephrase what he said.
Reinforcement	To support or strengthen learning.	Rephrasing a question that a student has answered; asking the student to elaborate.
Motivation	To provide stimuli that encourage students to inquire and to become responsible for their own learning.	What do you think will happen . . .? How can you test that theory?
Attitude Development	To move students in the direction of a favorable attitude toward science.	What effect do you think that will have on society? What do you think of that? How do you feel now?
Evaluation	To assess the progress of students.	Asking any of the evaluative questions.

Your Students' Behavior

Look at the collage of photographs (Figure 29). Notice what the students are doing (or not doing). How do students in your classroom behave? Do students in your class initiate behavior? (That is, do they ask questions without being motivated by you?) Or, do students predominantly respond to your behavior? Now in the photographs we can't tell what the students are saying, but we can focus on some of their nonverbal behavior (looking at another person, handling science materials, writing, etc.).

An approach we have found useful to answer questions about our students' behavior is to use the SCAN (Student Communication Analysis) System. It was developed to observe students while learning science.

The SCAN card is shown in Figure 30. It is organized into communicative and noncommunicative behaviors. For instance, within the communicative behavior section, a student is either initiating behavior, or responding to someone else's behavior. Do your students primarily initiate or respond? Using the instrument, while focusing on several students, will help you answer that question. Other questions you can ask are:

Is the behavior of students I'm observing instruction-related or non-instruction-related?

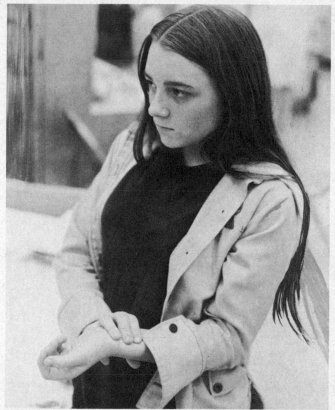

Figure 30.   SCAN card.[9]

	Communicative Behavior	Student 1	Student 2	Student 3	Student 4	Total

Initiating	1. *Instruction-Related Examples*
	     Asking a question.
	     Going to a learning
	          center.
	     Raising hand.
		1	1		111	5

	2. *Non-Instruction-Related Examples*
	     Throwing a spit ball.
	     Talking to partner
	          about ball game.
				11		2

Responding	1. *Instruction-Related Examples*
	     Answering a T. question.
	
	     Raising hand.
	     Working with equipment.
	     Looking at teacher.
		1111
		1111 1
			1111		1111	21

	2. *Non-Instruction-Related Examples* Getting hit by spit ball.				11	2
	Noncommunicative Behavior					
Active	*Examples* Playing Reading—nonclass Writing—nonclass Doodling Look Glance Adjusting apparel Grooming			1111 1111 111		13
Passive	*Examples* Head on desk Eyes closed Fixed stare			111		3

· Do students I work with initiate behavior verbally or nonverbally? Do they respond?

· What percentage of their behavior is noncommunicative?

For example, imagine a colleague has visited your class and recorded what a few of your students were doing using the SCAN. The results are shown on a SCAN observation card (Figure 30).

To use the SCAN card, you should focus on one student for five minutes and record a tally for each type of behavior observed. The predominant student behaviors for this class are

1. Responding to instruction-related—21 instances.
2. Noncommunicative, active—13 instances.
3. Initiating, instruction-related—4 instances.

You also can tell that one student (# 3) apparently is bored, while student 1 is the most responsive. Use the SCAN to find out how students in your class are doing.

A Classroom Checklist

Another interesting technique is to have one of your peers visit your classroom for two or three sessions and then complete the checklist shown in Figure 31. After the checklist has been completed, you and your colleague can go over the results.

Figure 31. Science classroom checklist.

Respond to each question by circling O (observed) or
N (not observed). Comments may be made.

A. Classroom Organization and Environment   Comments

1. Each student has a space of his own.   O   N
2. Desks and chair are arranged in rows and columns.   O   N
3. Room has interest or learning centers.   O   N
4. Areas are utilized to display student's work.   O   N
5. Students generally work as:
   large groups   O   N
   small groups   O   N
   individuals   O   N
6. Materials accessible to students.   O   N
7. Students seem to enjoy being in the room.   O   N
8. Students are involved in some planning activities.   O   N
9. Student work and learning
   organized through the use of:
   learning contracts   O   N
   textbooks   O   N
   teacher verbal directions   O   N
   workbooks   O   N
   student decisions   O   N
   other   O   N
10. Students seem free to move about
    and engage in various learning
    activities.   O   N
11. Science activities include:
    individual lab work   O   N
    outdoor field trips   O   N
    lessons that focus on
        processes of observing,
        classifying   O   N
12. Science activities include:
    project work for exploring
        interesting topics   O   N
    valuing activities   O   N
    group discussions   O   N
    inquiry demonstrations   O   N
    reports by students   O   N
    films, slides, tapes, and
        other media   O   N

B. Teacher-Student Interaction   Comments
     1. Teacher does most of talking.   O   N
     2. Teacher behavior is mostly verbal.   O   N
     3. Teacher uses variety of nonverbal behavior.   O   N
     4. Teacher moves about the room.   O   N
     5. Teacher accepts student ideas.   O   N
     6. Teacher generally directs
         student learning activities.   O   N
     7. Time is spent on nonscience activities.   O   N
     8. Children generally dependent on
         teacher for learning activities.   O   N
     9. Students participate freely in class discussions.   O   N
    10. Teacher maintains close physical contact with students.   O   N
    11. Evidence of student-student interaction.   O   N
    12. Students initiate some of their activities.   O   N
    13. Creative behavior is encouraged.   O   N
    14. Students have freedom to explore
         science topics based on his interest.   O   N
    15. Skepticism is valued.   O   N
    16. Originality of expression is encouraged.   O   N
    17. Various forms of student communication
         permeate the classroom.   O   N

Some Ways To Become More Humanistic

Every day of teaching is a day of choice. We choose to progress and try untried activities to see how they work, or we choose to keep working on the tried and true. Each day we make choices about something a lot more personal to us than any science activity: we choose to grow and become better than we've ever been, or to remain at the same point for another day, another week . . . another year.

There are some things we can put into practice not tomorrow or next Tuesday or next year but today that will help us to become more humanistic. We would like to share 15 of them with you. Here are some ways to shake out those rainy day blues and let the sun shine in!

Let the sun shine in!

1. Get alone each day for a least 10 minutes for reflection and meditation about where you are going as a teacher.
2. Ask yourself: How do I feel about teaching? Am I accomplishing what I want? What are my goals as a teacher?
3. Keep an ongoing *teacher log* in which you write observations of, reactions to, and feelings about your work as a teacher.
4. Visit other teachers in your school. How do they teach? Find out what you can learn from them. Trade teaching positions from time to time once in a while.
5. Play some music in your class instead of doing what you planned. Select music you like! On another day let the children bring in music they like.
6. Once in a while let the children in your class plan the entire day or what they do during science. Later let the class be partly responsible for planning the entire year.
7. Structure your language from time to time on separate days. Try the following:
   a. Speak only in the first person when talking with your students.
   b. Confine your teaching to asking questions: do not lecture.
   c. Confine your teaching to responding to student questions.
   d. Speak for intervals no longer than 30 seconds: the students must pick it up when time is up.
8. Incorporate some alternatives to testing and evaluation.
   a. Use student-made tests and take them yourself.
   b. Use open-book tests.
   c. Use take-home tests.
   d. Use mastery (not mystery) tests which the students can take until they master the material.
   e. Allow students to take tests in groups of three. Individual grades will be based on group results.
   f. Place much less emphasis on testing by having test scores count less than one-fourth of student's grade.
   g. Use tests only for students' self-evaluation. Tests are not graded A, B, C, but instead are used as feedback to the students.
   h. Have students indicate what they learned through art-work, music, or poetry.
   i. Make up your own alternatives.
9. Experiment with so-called free time. Let the students do as they wish during this time.
10. Try to "teach" something to your students that you know nothing about. How do you feel about this?
11. Tape record your classes and make it a habit to listen to yourself.
12. Do value clarifying activities (Chapter 8) once per week. Involve your students in deciding upon topics and activities you will do.
13. Have a cook-in. Students should bring a food they actually made to class. Insist on recipes. Perhaps it will turn into a party.
14. Conduct feedback or rap sessions with your students. This can be done during school or as a volunteer activity after school.

Use other evaluation techniques suggested earlier in this chapter.

15. Record your reactions and feelings in your *teacher log* each time you try one of these.

Workshops

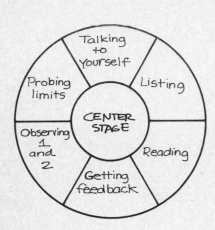

Center Stage

You're teaching your fifth grade class, and a sixth grade teacher in your school calls you out into the hall. She complains to you that you are not dealing with your students in a professional manner. She says you are too personal with the students and your planning is inadequate and then raises moral and ethical questions with you.

How do you feel about what she has said? What do you say to her? What are you going to do? Choose one of the personality roles and play it out.[10]

Level 1.   I am the most important person in the world. Everyone exists to serve my needs. I am never guilty; if I make a mistake, it is someone else's fault.

Level 2.   I am not satisfied with myself and am constantly looking for people I can emulate who appear to be better than I am. Sometimes I am really aware of my feelings and it scares me. I jump on bandwagons and will always be a winner.

Level 3.   I am a humanized person. I have empathy for others. I view others as real human beings. I may appear ego-centered, but I am not. I can laugh and cry at myself. I am in touch with my feelings. I can experience real guilt, love, hate.

Listing

Make a list of practices you think are important in teaching, such as "each student decides how much time he should spend on a topic." Exhange lists with someone else. How do you compare? Discuss your differences and the things you agree on.

Talking to Yourself

Write your own interview by responding to each of the following questions.

1. How do you feel about teaching?
2. What do you think is the major contribution you make as a teacher, to your students?
3. How would you or do you operate your classroom? Consider choosing content, activities, organization of activities, grading, and evaluation.
4. How do you feel about using student feedback in teaching?
5. To what extent do you think students should have the freedom to choose what they learn?

Reading

Read the first chapter, "A Sixth Grade Teacher Experiments," in Carl Rogers's book, *Freedom To Learn.* How do you feel about this teacher? To what degree, do you think, has she become aware of her realness? Where are You?

Probing Limits

What things in your environment limit you? Get together with a small group and talk about this.

Observing 1

Be the student. If you are a student in a course, study the teacher's questioning skill by coding the teacher's questions using an instrument shown in the chapter. Summarize your results and share them with the teacher. How did he feel? How did you feel?

Observing 2

Make up a scheme for observing science teaching. Visit a classroom and observe the teacher using the instrument you have developed.

Getting Feedback

Teach a lesson to a group of children or peers and obtain some form of student feedback using one of the techniques suggested in the chapter. How do you feel about doing this right now? If you do this workshop discuss with someone else how you reacted to the students' responses.

Reference Shelf

BECOMING MORE HUMANISTIC

A Conversation With Real Teachers

We ended the previous chapters by sharing conversations we had with each other. In this chapter we wish to share conversations we had with Anita Bradford, Carolyn Hunsicker, and Peggy Daniel. We selected these people because of their attitudes, ideas, and enthusiasm. In fact, when asked how they felt about teaching, they all responded with "I love it!" Because we know these teachers are people who do the hard work of teaching and yet still enjoy it by going beyond loving, we thought you would like to meet them.

Anita works with six- to ten-year-old children and has been teaching for six years.

Anita.

Carolyn has been teaching for 14 years and presently is a sixth grade math and science teacher.

Carolyn.

Peggy is a seventh grade science teacher and has been teaching for 12 years.

Peggy.

Jack: How do you feel about teaching?

Anita: I love teaching. Most of the time it doesn't seem like work. Because I remember elementary school so vividly and have a great deal of sensitivity about childhood, it makes me feel great to provide an environment for kids where they feel safe and secure and where learning is fun.

Carolyn: Teaching is what it is all about! *I love it*—find it challenging, exciting, and refreshing even with its frustrating moments and its TGIF weeks! Where else are we continuously on stage and have such a captive audience? What frightening power over so many lives!

Peggy: I feel teaching is a challenging profession: each day is a new day and each student is unique! I think I find "reward" in the friendships I make with students, many of which survive the years! It is frustrating if I choose to let it be frustrating. For me frustration is a signal that something I am doing in my classroom and with my students is not *our* thing. And too, the way I choose to go about conducting my classroom time requires quite a lot of outside time, and there never seems to be enough of that. Frequently, there is the search for inexpensive, free "stuff" we need—and always something I feel I need to read for my benefit. It is boring in the same way it is frustrating. Thinking back, I believe I was becoming bored with teaching at the

same time I was experiencing my greatest period of frustration. I was still very "traditional," and students were bored and frustrated too. Life itself is exciting and school should be likewise.

Joe: You've indicated some of your feelings about teaching. I'm wondering about your beliefs. What are some of your strongest beliefs about teaching?

Anita: I believe in creating an environment in which children are given diverse opportunities for learning in an atmosphere of loving acceptance.

Carolyn: Learning takes place in a variety of ways. A teacher need not be the central figure. A majority of learning taking place in my classroom is peer-group oriented. I think we as teachers need to learn to trust children's world of curiosity, imagination, feelings, and exploration. I also believe there is a need for a balance between what is called traditional and the open classroom approach.

Peggy: I believe there are varieties of ways in which people learn and teaching should present information in a variety of ways. I believe people learn best that thing they are most interested in, in an environment of openness and trust. Most of the choice for what one learns and how he learns it should be the individual's responsibility. If he is unable to assume that responsibility, I as a teacher should be able to structure the work so he can learn to assume responsibility. I do a lot of that too. I feel as a teacher I should meet each student (idealistically, *most* students realistically) where he is; and by whatever means necessary, take him as far as he is willing to go.

Jack: Let's go from philosophy to practice. How do each of you operate your classroom. How do you choose content, select activities, and evaluate students?

Anita: We choose content with the entire Early Learning faculty. Most academic work is done in small groups. We usually have several projects going on at any one time in addition to basic grouping for math, reading, spelling, etc. The projects have included publishing a class newspaper, a lengthy social studies–science unit on endangered species, and comparison of Greek and Norse mythology. Evaluation is primarily informal and based on the observed application of new skills.

Carolyn: I have a laboratory-oriented classroom. I call it child-centered as opposed to teacher-directed. Our program is individualized, and the teacher serves as a catalyst or facilitator. Content activities are selected by the students based on basic content selected by the teacher.

Peggy: Normally I teach three regular sections with two accelerated. I comply with the "guidelines" in that I spend the first period of time with the methods of science, the metric system of measurement (no conversions please!), laboratory equipment, record-keeping practices, and through all of this I try to focus ever-present change. I provide about 60 to 70 percent lab activities for learning all this. If time will not permit all I would like for students to accomplish, I ask them to vote on which activities we will pursue. We do an awful lot of voting in my classes.

I think kids in this day and time need to make many decisions, and school rarely affords them this opportunity. I am really flexible; and if students dream up an experiment about something we are studying or that they are interested in, I can go all out to help round up equipment and so on.

I go all out for student evaluation of their own work: that is, at a check-up time, each student gives himself a grade for his work. Most of these grades are lab work grades. I stress that each student knows just how much work he did

and what he learned as a result and he should best know what grade he deserves. Yes, sometimes a student assigns himself a grade he simply *wants* and perhaps doesn't deserve. But then students do have those kinds of needs too, don't they? The more academically capable students frequently have more of those needs.

We rarely do written questions from the book as this leads to a lot of "copying." I usually assign various questions to various groups for presentation to the class. I grade all "test" work. Tests are a variety of things, sometimes an objective one (open book), an essay question or two, analyzing pictures, doing a special lab, graphing, using data I supply, and so on. I use value lines in addition to the student assignment of number grades to lab work. I urge a student to be totally honest about the value of the experience he had and, on a scale from 0 to 10, place his "X" for the value he considered the experience to be. It need not match the self-assigned number grade at all.

And one other thing, this year I'm trying to do a lot of peer teaching. We work toward doing an exciting job of presenting whatever it is in such a way that nobody will be bored. This normally results in quite a bit of creativity.

Joe: Do you use any special techniques or strategies in your classroom?

Anita: I don't use contracts in teaching science. I do use learning centers, but the children prefer group process activities to my learning centers. I'm working on a design to see if I can get a more varied set of activities.

Carolyn: I use a variety of techniques. I use contracts, learning centers, individual learning packets (minicenters), independent study, student-prepared audiovisual media, "buddy" system learning, and outdoor classwork.

Peggy: I've already mentioned my evaluation techniques and heavy involvement of my students in making choices as to what they will learn. I'd like to mention how we make use of the outdoors. About three years ago we put in an outdoor environmental lab in a large space between two of the wings of our junior high school building. We built "hills," and the student council donated a fiberglass pond and a windmill for circulating water in the pond. We have experimental plots of natural grasses, crops of certain wildflowers for feedstuff for migrating birds, a bird-feeding station, and free trees secured from the State Department of Conservation. Now the main way I use the outdoor lab is for slow achievers: many of them get quite turned on with a regular "chore" in the outdoor lab, and we contract for a grade for doing such a chore regularly and reasonably well. It is a boon to poor readers who may choose to do some gardening or grounds keeping instead of reading the text. These students still perform laboratory work with partners.

Jack: I'd like to ask two questions. First, what is the major contribution that you as a teacher make? Second, have you made many changes in your teaching recently?

Anita: I'm not sure about your first question. I think that it's the fact that I am open and honest as I can be and the children feel they can trust me. As far as changes, I've become less and less caught in the "teacher" role. I've become comfortable as an adult person who lives with younger people. This has meant that I can allow them to take the responsibility for their own participation in their own learning and also allow them the natural consequences of their own acts.

Carolyn: I think the most important contribution I make as a teacher is that I am a human being, and then a teacher. I enjoy life, feel emotions, love my work and my students, and I am not afraid to show it. I am not plastic. Recently, I have made few changes. But one point is important to make. I have been willing to give up known, workable, old standby techniques and methods to venture down some untraveled paths (with no guarantees). If I'm wrong, I try something else. I'm willing to try anything to give a child a measure and taste of success!

Peggy: My ability to listen to them and to reflect back to them what I have heard them say—this in the realm of the personal relationship we establish as a result of the openness I try to provide in my classroom. Many students have told me that they learn early that I truly care about them as young people. And I truly do. I guess really I hope that the major contribution I make is to model patience and trust and concern. I do this in the academic area, and they seek me out for the personal needs they have. (Incidentally, this includes loaning money, and I have never lost a dime!) Most recently in this school year,

I have incorporated various value clarification strategies into my classroom routine. These have met with tremendous success.

Joe: I'd like to ask two other questions. First, do you use student feedback to check up on your teaching? And second, do you think your kids like science after having you as a teacher?

Anita: I have constant verbal feedback from the children. They are not at all hesitant about letting me know what they feel about what we are doing.

Yes! Kids like science because science activities offer more opportunities for active involvement than any other curriculum area.

[We doubt that it's just the curriculum that turns Anita's children on!]

Carolyn: Student feedback flows freely. Although I don't use any special forms, feedback emerges from open-ended discussions, laboratory work, and planned student feedback. The key is listening to it!

Most generally, the students like science. Not all are budding scientists or have the same feelings for math and science as I do; however, they do have the opportunity to develop an awareness and sensitivity to things around them.

Peggy: Yes, I do use student feedback to check up on my teaching. I do not have a "standard" form. Most generally I ditto open-ended sentences such as:

"The best things about Chapter 6 were . . . "
"The worst things about Chapter 6 were . . . "
"Mrs. Daniel should have . . . "
"In the future I wish Mrs. Daniel would . . . "

These may be for a chapter, a test, a film, a field trip, etc. The last one listed above brings in a multitude of responses—pro and con—and if many say the same thing, I really do regroup. I usually amalgamate the critiques and return the composite to the class for their analysis and observations. I do try to operate for the benefit of the majority.

Do they like science? Yes, the majority of them do. I equate it with their being able to like me as a person and to having a very good time in my classes—*doing* a lot of science!

Jack: Anita, Carolyn, and Peggy, we thank you for sharing your views on teaching with us. I'm sure our readers appreciate your enthusiasm and insight into teaching children.

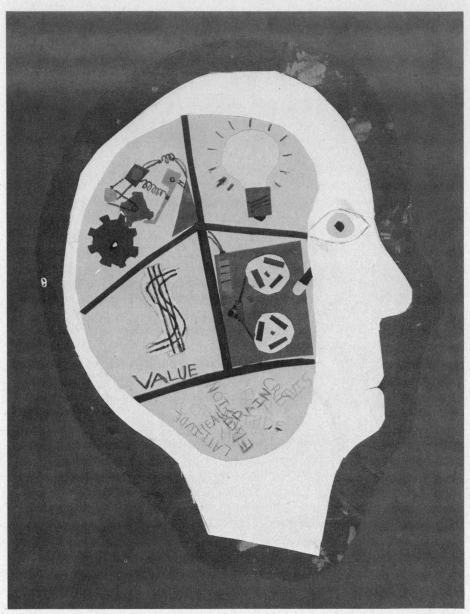

Cut and pasted construction paper.

6
From
Teacher Stuff
to Kid Stuff

The heart and soul of this book is tied up in the last three chapters. We must go beyond loving children to the extent that we channel our love into actions. Maxine Greene, writing about the existential teacher, puts it this way: "The point is that mere expression of values means nothing; the expression must be acted on and realized."[1]

We have chosen to direct our actions into three processes, *discovering, valuing,* and *exploring,* and have included an array of learning activities that give meaning and importance to each of these processes. Together we see these processes related to each other to form a meaningful and holistic view of learning and science. In this chapter we will discuss the rationale for the three processes, how they are related to science, and how they can be applied to turn your classroom into an environment full of wonder, curiosity, freedom, and inquiry. In short, into a humanistic learning environment.

Science Is a Verb

"Only in relatively recent years, has the view of science as a process been exploited. This is to say, science as an activity rather than a thing is a view which is of recent origin and is, as yet, not a common mode of operation in teaching at any level. Considering science as an activity, that which one does, raises a question regarding the essence or nature of science. Is science a noun or a verb?"[2]

In discussing this question, Morgan suggests that "knowledge is initially derived from observation. This is the first stage or step in sciencing and is a basic or fundamental process of science."[3] This sensory-motor stage is followed by a second sciencing stage of classification, which is followed by a cognitive and experimental stage. Science is a doing activity, and if it was considered to be a verb, the teaching of science could be devoted to the pursuit of a child's natural sensory curiosity, thus leading to a fuller understanding of science and life.

Learning science should be fun, and should involve a very wide range of experiences, including everything from what some have labeled "messing about"[4] activities to very carefully structured encounters typifying the rational model of science. We know that the history of science reveals science as not always proceeding in a very rational, logical, orderly way; but because of the reporting process (magazines, journals), much of the irrational, intuitive, and playful aspects became lost.

Many books portray science as a collection of facts, or a "body of knowledge" to be learned. Unfortunately, the impression we may leave with children is that science is nothing more than learning facts and theories. There is much room in science for intuitive, hypothetical, playful, and imaginative forms of learning.

Recent discoveries about how we think and learn suggest that the human brain has two distinct kinds of thinking capabilities: the intuitive and aesthetic, and the logical and rational. Because these abilities seem to originate in different halves of the brain (left and right cerebral hemispheres), they are sometimes referred to as left-handed thinking and right-handed thinking (Figure 32).

The activities that appear in Chapters 7, 8, and 9 include a blend of both left- and right-handed encounters and suggestions for you to use with children. Chapters 7 and 9 include a preponderance of right-handed activities, while Chapter 8 emphasizes left-handed knowing.

SCIENCING

analyzing
MEASURING
CLASSIFYING
VERIFYING
writing
ORDERING
USING
SPACE-
TIME
rela-
tion-
ships
REASONING
planning
READING

feeling
dreaming
LOVING
VALUING
DRAWING
IM-
AG-
IN-
ing
singing
INVENTING
INTEGRATING
fantasizing

RIGHT-HANDED THINKING (controlled by the LEFT cerebral hemisphere)

LEFT-HANDED THINKING (controlled by the RIGHT cerebral hemisphere)

Bob Samples, while discussing the rationale for the integration of left and right-handed knowing, notes that,

> Our early excursions into building these constructs suggest that metaphoric or intuitive maturity is quite available to us all. It involves, in the words of Robert Oppenheimer, getting our mind to work in ways like the children playing in the streets. Humans cannot escape intellectual maturity. If Bruner and Piaget are right, then exposure to a technocratic society will all but insure it. So to focus on the development of intuitive and metaphoric maturity is not to discredit the rational-logical mind. Instead the purpose is to celebrate the union of our minds. To wed the functions of intuition and metaphor with the functions of logic and rationality and create the potential for the synergic mind.[5]

Figure 32. Left and right mind functions.

Teacher and Child Together

One of the most beautiful things we can imagine is a child and a teacher learning together. We feel very strongly that the humanistic teacher cares about this and does his or her very best to help this happen. We envision this occurring as the result of the interaction of teacher and child within three processes of learning and sciencing. We refer to these processes as:

Discovering

Valuing

Exploring

Each of these three processes forms the basis of a chapter in the Kid Stuff portion of the book.

Discovering

Learning and sciencing start with an awareness of our environment. For children, discovering occurs when they come in contact with their surroundings through activities involving the senses. The teacher's role is crucial in this stage. He or she brings the child in contact with the environment by helping the child discover his sensory powers and expanding his inquiry processes. This is an important role. Without the guidance, planning, and hand of the teacher, learning could proceed aimlessly. Thus, in the discovering process, the child and teacher work hand in hand to discover their environment.

The discovering process is dominated by psychomotor activities of touch, taste, hearing, smell, and sight. It represents the observational and perceptual aspect of learning. The chapter on discovering (7) contains an array of learning activities aimed at discovery learning. Sensory perception, process activities (observing, classifying, measuring, and inferring), and inquiry problems dominate the chapter.

The winner in teaching we referred to in Chapter 1 provides a classroom environment rich in materials, objects, and activities that bring the child in contact with reality. Such a teacher probably will have to

1.   scrounge for learning materials,
2.   develop good small group teaching techniques,
3.   help children ask questions about their surroundings, and
4.   develop a repertoire of discovery learning activities.

Many of the activities described in Chapter 7 are designed so that you become a leader, a catalyst, and a facilitator of the activity process.

Valuing

The second stage or process we have chosen to emphasize is that of valuing. We cannot ignore children's feelings, emotions, and attitudes as they learn science. We've ignored affective development at the elementary, secondary, and collegiate levels of education far too long.

The valuing process involves affective qualitites of learning, such as feelings, emotions, and attitudes. Valuing activities, such as those in Chapter 8,

represent the other side or the left hand of learning. Valuing activities turn science into a humanistic endeavor, helping children realize that science discoveries and explanations are made by people, that science is the result of human efforts, and that it is not a sterile body of facts and knowledge.

The teacher's role in valuing activities is again that of a facilitator. Small group learning is encouraged if you utilize the valuing activities in this book. Sometimes you may actually be in the group, while at other times your children will be valuing on their own.

There may be several groups working on the same activity, or you may work with the entire class together. Warm-up, awareness, and value lesson activities are included, and we feel very strongly that you and your children should engage in these activities together.

Exploring

The exploring process is in the cognitive realm of learning. It represents the research and study associated with learning and sciencing, and occurs when a child has reached a sufficiently high level of awareness, interest, and motivation to pursue science topics in depth. Now that you and your students have made a discovery and placed a human value or worth on discovery, you face the challenge of continued exploration. Activities in Chapter 9 are designed to provide you with ideas for student projects, field trips, student experiments, and learning centers.

We must be cautious and not assume that our role as the teacher is to *give* the student the knowledge and information that is typically associated with exploring. Kahlil Gibran, in his book *The Prophet,* says it eloquently.

> *No man can reveal to you ought*
> *but that which already lies half*
> *asleep in the dawning of your*
> *knowledge.*
> *The teacher who walks in the*
> *shadow of the temple, among his*
> *followers, gives not of his wisdom*
> *but rather of his faith and his*
> *lovingness.*
> *If he is indeed wise he*
> *does not bid you enter the*
> *house of his wisdom, but rather*
> *leads you to the threshold*
> *of your own mind.*
> *The astronomer may speak to*
> *you of his understanding of*
> *space, but he cannot give*
> *you his understanding.* [6]

Children doing exploring activities will be engaged in a variety of learning modes. Sometimes they will be working alone (reading, making a collection of rocks, working in a learning center, writing a story), or working with one or two other children (doing an experiment, going on a mini field trip), or sometimes cooperating with the entire class (discussing the results of an experiment, mini field trip, or something they read about).

Because of this variety, your role will be multifaceted. You will have to spend much time setting up learning centers, making sure materials are on hand for experiments, arranging for small groups to go outside the classroom, questioning and clarifying what your children are doing, keeping up with their progress, and so forth. You and your children will find yourselves exploring in a wide variety of learning situations.

Using the Kid Stuff

It is our hope that the humanistic framework developed in the Teacher Stuff portion of this book can be put into practice and facilitated by the next three chapters, as well as with the many suggestions made earlier. The Kid Stuff chapters are a resource of ideas and approaches to science teaching for you to pick, choose, arrange, rearrange, and modify to suit the needs of your students and you.

A humanistic classroom does not mean that students are turned loose to do as they want, or that everything is individualized, or that the classroom is organized around learning centers. We believe that a humanistic classroom occurs when a teacher and the thirty or so children in his or her classroom face each other and deal with each other as human beings. It begins with acceptance of each other and this includes the provision that alternatives for learning will be available. The environment grows and becomes rich with ideas, people, and things to do. A humanistic environment is more psychological than physical, and is based on action rather than appearance.

To help you pick and choose from the activities in the second part of the book, we have organized the ideas into an overview chart (Table 10). We invite you to glance at some of the activities using the chart as a guide. There are 81 activities in the three chapters, but we have listed the names of only a few to give you a flavor of what's to come. As you do activities, we have included space for you to write your impressions on the chart, or you could make notes in a *teacher log*, if you keep one.

Table 10.

Kid Stuff

Activities	Purpose	Sample Activity Titles	Space for Evaluation of Activities You Do
Discovering			
Sensory awareness	*To help children discover their environment using all of their senses (sight, hearing, touch, taste, smell).*	*What Do You Feel?*	
Science processes	*Activities to help children learn basic processes of observation, classifying, measuring, and inferring.*	*The Old Fossil* *Body Lengths* *The Mealworm Race* *The Footprint Puzzle*	
Science inquiry problems	*Staging discrepant events to get children motivated to ask questions and inquire.*	*Moon Crater Inquiry* *Muddy Current Inquiry*	
Valuing			
Warm-ups	*Designed to help you and your students loosen up to learning.*	*Know Your Rock* *Blind Walk* *Fantasy Trips*	
Awareness encounters	*Activities designed to help children become aware of their feelings and attitudes toward nature.*	*Sounds of Nature* *Art and the Environment* *Body Mapping*	
Value lessons	*Activities designed to help you and students clarify attitudes and feelings.*	*Meet the Scientist* *Role Playing*	
Exploring			
Student projects	*Designed for independent and small group work. Emphasizing creative behavior.*	*The Egg Drop* *Animal Tracks*	
Field trips	*Suggestions and ideas for mini field trips and yellow page field trips.*	*20 Mini Trips*	
Student experiments	*Learning Activities to teach children processes of model building, controlling variables, hypothesizing, and designing experiments.*	*Inquiry Boxes* *The Rocket*	
Learning centers	*Activities for independent or small groups that are based on an integration of process of discovering, valuing, exploring.*	*Dig the Earth*	

Each chapter that follows contains suggestions in terms of materials, approach, and modification for each activity described. We hope that your classroom and the lives of your students will be enriched as you apply these activities and processes. The challenge is for you to put them into action.

Workshops

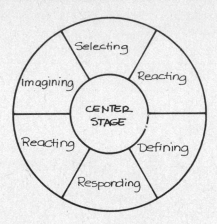

Center Stage

You have just returned from a conference on humanistic learning and are anxious to try out some of the activities that you encountered there. While you are outside with your third grade class doing some of the activities, your supervisor observes some of your children leading others around blindfolded. She immediately comes to you and asks why during science children are doing such things. She also wants to know why you are outside.

What are your feelings about her remarks? How will you answer your supervisor? What do your students think?

By the way, what were your students doing walking around blindfolded?

Selecting

Select any activity from Chapters 7, 8, or 9. Do the activity with a small group of children or peers. Select one of the forms of feedback suggested in Chapter 5 to evaluate the lesson.

Reacting

A student comes up to you and says, "Hey teach, let me do an experiment like Jeff, too!" (He's been a discipline problem all year—you know the type.) What would you do?

Responding

A fifth grade boy has made a "discovery." He comes up to you after science time at 2:10 and says, "Mr. Glick, I think the gerbil is having her babies. I'd like [valuing] to stay to find out how she does it [exploring]." What would your comment be in each of the following situations?

1. You are a teacher in a self-contained classroom and the thing the class usually does at 2:10 is math.
2. The fifth grade schedule is departmentalized, and your children are supposed to be going across the hall for math class.

Reacting

Suppose a second grade child brings a frog which was found in her backyard and asks, "Miss Zing, where do frogs come from?" How would you handle this?

Imagining

Imagine these scenes in two different classrooms.

Scene 1. After taking a blind walk through the school nature trail, each of the students is making a collage using materials experienced during the walk.

Scene 2. Students are sitting at their desks writing a story about bees in a meadow that they were going to visit today but could not because it was raining.

Which learning activity would you choose to do with a group of children?

Defining

Write your definition of science and science teaching. How do they differ? Compare your response with those of others.

Reference Shelf

FROM TEACHER STUFF TO KID STUFF

Conversation

Joe: This is really a key chapter in the book. It's a transition that gets us into actual science activities teachers can use.

Jack: I'm glad we brought out the idea of science as a verb, or should I start using the term "sciencing." "Sciencing" is really a good term: it sets the stage for the next chapters very nicely.

Joe: The idea of science conveying action is something that many people really don't comprehend.

Jack: You're right. I hope that reducing the learn- ing of science to three processes of discovering, valuing, and exploring will provide teachers with a usable framework.

Joe: I think it will. Keeping science simple, rather than complex, and viewing it as a process will make it more enjoyable.

Jack: If people could just keep that it mind, the children in our schools would be so much better off. That includes teachers, administrators, and college professors.

Joe: You're getting louder.

Kid Stuff:

Hugging a Tree
and
80 Other Science Activities
for Children

Cut and pasted construction paper.

*The most remarkable discovery
made by scientists is science itself.* [1]
J. Bronowski

7
Discovering

S cience is a process of uncovering and attempting to explain the mysteries and unknowns of the natural world. Science for children should nurture their curiosity about the world, their love for asking questions, and their playfulness.

Science starts with observations of reality, and for children this is the world that surrounds them. Young children have a natural curiosity to know things about the environment around them. They constantly ask questions and make statements such as: ''How does that feel?'' ''Let me touch that!'' ''Look at that fossil!'' ''What does that taste like?'' ''What is that loud roaring sound?'' Their inquiry starts with their own experiences, their own observations.

Discovering, as used in this book, is the awareness stage for the process of sciencing. We become aware of rocks, plants, the stars, animals, water, solid, flowers, objects, and *ourselves* through sensory experiences.

In this chapter, which we have titled ''Discovering,'' emphasis will be placed on activities for children which focus on sensory experiences. We will expand and enlarge the process of discovering to include science process activities and science inquiry problems also. The humanistic science classroom should feature a variety of ways of knowing, learning science, and discovering.

A child discovers her heartbeat.

Sensory Awareness

In the classroom children should be given the freedom to explore their environment using all of their senses. Unfortunately we ignore most of our senses (except sight and hearing) in planning lessons. The senses are the basis for scientific observations. (Telescopes, microscopes, and seismographs are instruments that simply extend the power of our basic senses.) Don't be fooled by the gadgets of science; science starts with us. Using our senses is a skill that can be learned and improved upon with experience and opportunity to "sense." Our science classrooms should be rich with sensory awareness activities. Here are a few.

Sight

Objects in a Box

Give the students a box (shoe box) with about five objects attached to the inside (a variety of rocks, or leaves, or miscellaneous objects). Cover the box with clear plastic wrap. Tell the children to describe the objects as best they can without touching or doing anything to the box.

For younger children, make the differences from one object to the next obvious (in color, size, shape); for older children, you can make the differences more subtle, which will heighten their discriminatory abilities.

Shadows

Using an overhead projector, project the shadows of a variety of objects on a screen. Do not let the children see the objects you are projecting. A cardboard screen mounted on the front of the projector will allow you to work behind it so they do not see the objects.

Use a variety of two- and three-dimensional objects, such as paper cutouts of squares, triangles, circles, ellipses, and rectangles, and solids, such as cubes, spheres, ellipsoids, rectangular solids, and cylinders.

You can turn this into a game by asking the children if they can guess what the object is by its shadow. This is not that easy. Imagine a cylinder placed on the projector so that it projects a circle. It could be a circle, sphere, or cylinder.

Touch

What Do You Feel?

This activity should be done blindfolded, or with eyes closed. Another alternative is to have "reach-in boxes" (feely boxes) in which the children put one or both hands in the ends of a box to feel the objects inside.

Give the children a collection of objects and ask them to describe each object by just touching (and nothing else). Give the students time to touch and explore each object. Now select one student and have him describe one object by giving one observation at a time. At the end of each observation, ask the other children if they can identify (still blindfolded) the object. If not, ask for another observation and so forth until one student can identify the object. Select another student and repeat the procedure. This not only helps the child with his sense of touch but with his ability to communicate.

Hearing

A Sound Walk[2]

Divide a piece of paper in half. On one side write "Man-Made Sounds," and on the other side write "Natural Sounds." Go for a walk on the school grounds for about fifteen minutes. Listen to the sounds around you. As you hear them, write them on your paper under "Man-Made Sounds" or "Natural Sounds."

Compare your list with someone else's list. What did both of you hear? What did you hear that the other one didn't hear? Were there any sounds that you heard that made it hard for you to decide which category they belonged in? If so, what?

The amount of time you allow for the sound walk will depend on the interest level of the children and the time you have available. It is important to allow enough time so that the children will have an opportunity to listen for quiet, subtle sounds as well as louder ones.

Taste

Observation involving taste should be done with caution. In general you should inform students *not* to taste things unless they are told to do so.

The Taste Test

Obtain a variety of food products, such as three brands of cola, three brands of peanut butter, three brands of chewing gum (same flavor), and three brands potato chips. Have the children pair off and provide small quantities of the food products. One member of the pair records the other member's reactions on separate data cards like the one in Figure 33. Tell the children to follow these rules:

1. Use small amounts of the product and the same amount each time.
2. Rinse out your mouth with water after each taste.
3. Try each food at least twice.
4. Blindfold the taster.

Instead of working in pairs, individual students may want to study the taste preferences of a large number of people.

Figure 33. Taste test data card.

TASTER_____ PRODUCT _____
TASTER'S FAVORITE BRAND _____
BRANDS USED_____

	TRIAL 1			TRIAL 2	
	IDENTIFIED			IDENTIFIED	
BRAND	YES	NO	BRAND	YES	NO

Smell

Observations done involving smell should also be done with caution. To smell an object, tell the child to fan his hand to and from his nose and to breathe normally. Some odors, such as those from acids, can be very harmful. This procedure will eliminate the problem.

"What Is It?" Box

Place an object in a small box which also contains an onion or mothball, or a small piece of cloth soaked in ammonia, vinegar, or perfume. Cut a hole in the box, but make sure there is no way for the students to see the object.

Let the students decide what is in the box by smelling. Repeat the activity with other objects. Do not be surprised if some students identify the objects as something very different.

An additional activity that you can do is to have the students determine how far away the odor can be detected. Are all objects detected at the same distance? Do all students smell things in the same way?

Mixing the Senses

This activity is designed to expand the student's awareness by using his senses with a "freshness that is not inhibited by accurate perception."[3] Table 11 is designed to encourage experiences that "mix the senses."

Give each child a copy of the table, or project one on a screen. Also, the activity is enriched if you have the objects, or pictures of the objects to use. You should be careful to keep the activity one of mixed senses. If you have something soft for the students to touch, make sure they describe what soft looks like, not what the object looks like.

After the children have the chart available to them, focus on one block at a time, such as How does soft look? and ask individual children to verbalize their answer or response. Using other forms of expression, such as drawing, music, and nonverbal expressions, can further expand the students' awareness.

Table 11.

Mixed Sense Chart[4]

	Sight	Taste	Smell	Touch	Hearing
Sight		What does red taste like?	What does the sky smell like?	What do mountains feel like?	What does blue sound like?
Taste	How does sour look?		What does sweet smell like?	How does bitter feel?	What does ice cream sound like?
Smell	What does the smell of rain look like?	How does perfume taste?		What do the smells of dinner cooking feel like?	What does the smell of soap sound like?
Touch	How does soft look?	What does rough rock taste like?	How does silky smell?		What does fur sound like?
Hearing	How does a whisper look?	What does laughing taste like?	What does barking smell like?	How does a siren feel?	

Science Process Activities

Getting children involved in science at the discovering level basically entails two things:

1. I experience.
2. I wonder about it.

Remembering that experience involves all the senses, you should be emphasizing the process of observation when planning science experiences. The wonder and curiosity will come about when your children are given the chance to infer and predict and go beynd their observations.

It's fairly obvious from everyday experiences that the "basics" turn out to be the important attributes of people in various professions and occupations: running, throwing, hitting, catching for a baseball player; examining, recording, observing, operating for a doctor; timing, gripping, holding, swinging for a trapeze artist. Notice that these attributes are "doing" things; it's the same for scientists and for boys and girls who are learning about science. The list of doing things that we refer to as processes is as follows:

1. Observing
2. Classifying
3. Measuring
4. Inferring

Teaching science to children (especially in the early grades) should emphasize these processes. In this section we will make some suggestions as to the types of process learning activities. Our concern is to provide activities as models you can use to develop additional process experiences.

Observing

The Old Fossil (No, Not You)

Give each student a collection of about five fossils, such as those shown in Figure 34, which you've either found yourself or ordered from a wish book. Ask them to describe each object (you may or may not want to tell them they are fossils) using all of their senses. We've included the names of these fossils for

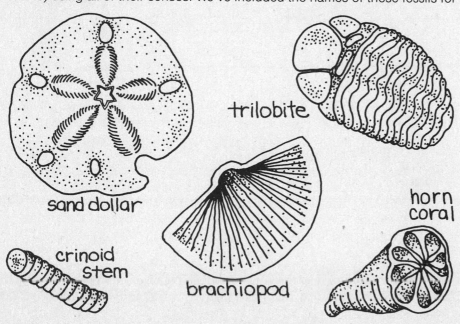

Figure 34. Fossils.

sand dollar

trilobite

crinoid stem

brachiopod

horn coral

your reference. Each child can produce a list of observations for each fossil on a piece of paper.

When you think they have had enough time to observe, you can then do any one of the following:

1. Have a child describe one fossil such as

It's kinda grey, has no legs; funny marks on the side...two eyes, and it's hard.

Ask other children to add observations to the list. You might get

Mine don't have no eyes, but it looks like it gots three legs. It is darker than grey, and smells like dirt.

Hey! This one is bigger than yours and has more ribs!

Yea... I wonder why?

Do this with each fossil. As the children describe each one, it might help to write their observations on the board. When the activity is complete, challenge the children by asking them if all senses were used to describe the fossils.

2. Place a set of about eight fossils in a cloth or paper sack. Only two or three should be those the children just observed. Challenge a student to reach in without looking and pick out a fossil he has observed. Turn this activity into a game by giving bags of fossils to groups of children. Points can be earned by selecting the known fossils.

3. Have the students pick one fossil; form five groups based on the fossil selected. The students can now compare the fossils that generally look alike. Raise the question, "How do the fossils that look alike differ?"

What You See Isn't Always What You Observe

Here's an activity that's a lot of fun and really tests a student's power of observation. Into several shoe boxes put the following objects:

Box # 1—Balloon filled with water
Box # 2—Ping-pong ball and small rock
Box # 3—Large washer and small ball
Box # 4—Nail and wooden spool
Box # 5—Toy car

Seal the boxes and tell the students not to open the boxes while challenging them to observe and find out as much as they can about the objects inside the box. Set them up at stations about the room and allow time for each child to play around with the boxes.

A lively discussion can revolve around questions such as: "What do the objects look like?" "How many objects are in each box?" "How big are they?" "Are they hard, soft, rough, smooth?"

This is an excellent activity to do if you are ready to work with your children on the process of making inferences. It's also a prelude to model building, which is an important process of science, because the child will be building a model or mental picture of the inside of the box.

More Things to Observe

Rocks
Leaves
Trees
Grass
Sugar cubes
Bugs
Insects
Large animals
Chemicals from home
     such as: toothpaste,
     baking powder,
     sugar, salt, flour
Seeds
Hands
Faces
Book covers
Microscopic plants and animals
Pictures in books, magazines
Photographs
The list goes on and on and on and on!

It Starts with Observation

Sciencing begins with observation . . . and we don't want you to go on too much further without giving this some thought. What *you* do is a key to how far your kids can go with learning. Observing brings the child into contact with something and may suggest how he can further investigate it. Here are some things that your children can do with what they have been observing:

- write a story about it
- draw a picture of it
- make models
- imagine where it came from
- create a poem about the object
- guess what it is
- find out more about it in a book or magazine
- make collections of the objects
- find other people who make collections of them
- go on a field trip to find more of them

You can also increase the power of their observations by giving them an opportunity to observe a variety of things and also asking them good questions. Ask questions that focus on such attributes as size, shape, and color. Here are some examples:

How large is it?
How many colors in your rock?
Did you see any other markings than these?
How many legs does your insect have?
How many babies did the guppy have?

Measuring

In spite of the complex image most of us have of science measurements, they are all encompassed in a simple list of three:

1. length measurements
2. mass measurements
3. time measurements

You can generate a good deal of excitement by placing about the room measuring problems such as those depicted in Figure 35.

Here are some suggestions if you wish to conduct separate lessons to help your children develop some basic skills in measuring.

Object Measuring

Give the students a collection of things, such as sand grains, small nails, cut strips of paper, coins, mealworms, or anything else you have on hand. In addition provide a small metric ruler and small piece of metric graph paper. Organize these items on a tray as shown in Figure 36 and give a tray to each student. Ask the children to measure the lengths of each object on the tray. Ask questions such as:

What was smallest?
What was length of smallest?
Did you all get the same length as Jimmy?

Figure 35. Measuring problems.

SHARK'S TEETH

How old do you think the shark's teeth are? Older than you?

microscope

HOW BIG ARE THE OBJECTS YOU SEE THE MICROSCOPE?

PLANT RICE

How far do you think each of these plants will grow in the next seven days?

MEASURING ants

HOW BIG IS THE BIGGEST ant?

JUMPING CRITTERS

HOW FAR DO THESE GRASSHOPPERS JUMP?

Figure 36. Object measuring tray.

Body Lengths

This measuring exercise is designed to help your children become aware of their own body dimensions. You should have available:

Bodies
Meter sticks
Small metric rulers
String
Butcher paper strips

Pose the problem, "How many different measurements can you make of your body?" Some of the things they may come up with are:

height
size of head
biceps
length of fingers
chest size
waist
thighs
fingernails
big toe
little toe

Encourage them to use a large sheet of butcher paper and report their results by cutting their body shape out and recording measurements. You can then play who's the biggest and smallest in the class.

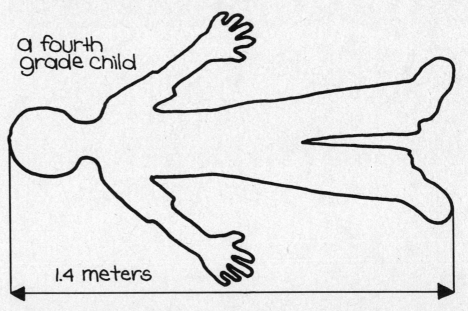

a fourth grade child

1.4 meters

"It Weighs 2 Paper Clips"

The concept of mass and how to measure the amount of "stuff" in an object is relatively simple. In this activity your students can build their own balance and use paper clips (or thumb tacks, pins, etc.) as their "standard" of measurement. Provide the students with a 20-centimeter strip of balsa wood, an 8-centimeter pin, 2 paper cups, piece of clay, milk carton, and a box of either paper clips, thumb tacks, or pins. Later provide them with objects to measure.

First ask them if they think they can make a balance. Give them time to "mess around" with the materials. Most of them will come up with something like the apparatus shown in Figure 37. The paper clip becomes a standard. All measurements are in terms of how many paper clips are required to balance the system. Give the students small objects such as:

acorns	beads
fossils	nuts
rocks	shells

Ask them how many paper clips it takes to balance any one of these. Results might take the form of:

	Acorn	*Shell*	*Nut*
Beth	9 clips	4 clips	13 clips
Anne	8 clips	5 clips	15 clips
John	10 clips	5 clips	13 clips
Jenny	8 clips	4 clips	12 clips
Jake	9 clips	4 clips	14 clips
Average	8.8 clips	4.4 clips	13.4 clips

Figure 37. A balance.

The Great Mealworm Race
This activity combines measuring length with the measuring time. Give each student the materials shown in Figure 38. Have the students measure the time it takes their mealworm to crawl to the end of a shoe box. After a few trials, conduct a class race in which you set all the boxes next to each other and see which mealworm crawls the farthest in 1 minute.

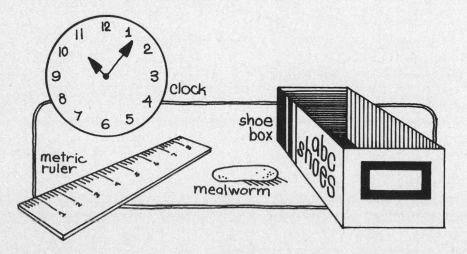

Figure 38. Materials for the Great Mealworm Race.

Measuring the Age of the Earth

This activity is designed to give the student a sense of how old the earth is and at the same time enable him to employ measuring skills.

Have the students work in small groups. Give each group a large piece of adding machine tape (5 meters long), crayons, and the following list of geological happenings:

Earthy Events

Events	Years Ago
Earth formed	5 billion
Age of moon rocks	4.5 billion
Oldest rocks	3.3 billion
First known plants (algae)	3.2 billion
First known animal (jellyfish)	1.2 billion
Paleozoic begins	600 million
First fossils	600 million
First fish	500 million
First land plants and air-breathing animals	440 million
First amphibians	400 million
Folding of North America	350 million
First reptiles	320 million
Great Ice Age in Southern Hemisphere	270 million
Mesozoic begins	225 million
First mammals	225 million
Continents begin drifting	180 million
Opening of South Atlantic Ocean	135 million
Beginning of Tertiary	70 million
First horse	60 million
Earliest elephants	40 million
First manlike animals	2 million
Beginning of Ice Age	1 million
Ice Age ends	10,000
Man on the moon	Less than 10

Roll out the tape and tape it to the floor. Allowing each meter to represent 1 billion years, have the students mark and label the positions of the "events" on the tape (see Figure 39). A discussion can follow concerning the age of the earth and the enormous amount of time since its beginning.

Figure 39. Adding machine tape used to measure the age of the earth.

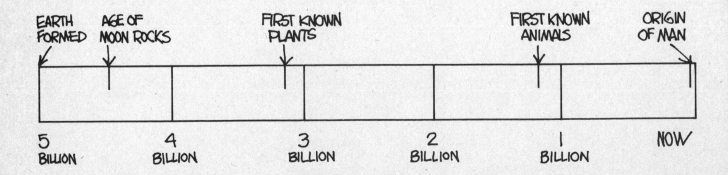

Classifying

Classifying is dependent on a child's observation skills. Classifying is a process in which a set of objects is divided into two or more subgroups on the basis of some observable property. Here are three classifying games that focus on observable properties as a basis for classifying.[5]

The Circles Game

This is a game for any number of players. Start with an assortment of objects (15 is a good number) grouped together in a circle drawn at the top of a large sheet of paper. The players think of a rule that will let them separate the objects into two or more subgroups (for example, things that are metal and things that are not metal). The objects are put into their new groups, and circles are drawn around each new group. These circles should then be labeled and connected to the first group by straight lines (see Figure 40).

The players continue to divide the objects into smaller and smaller subgroups, according to rules they think of, until each object is alone in a circle. Each subgroup should be labeled and connected to the one it comes from. Objects can include mixtures of objects, rocks, colored blocks, and so forth.

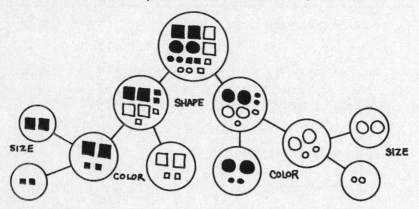

Figure 40. The circles game.

The Sorting Game

This is a game for two players. One child sorts all the objects in a box into two or more piles, according to some rule (see Figure 41). The other child tries to guess the rule and should be able to give a name to each pile. A variation might be for the children to keep track of their rules to see who has thought of the most rules to divide his objects by.

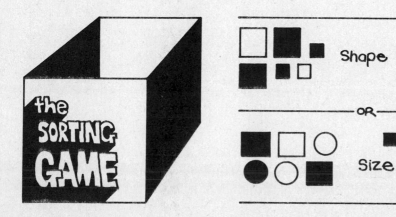

Figure 41. The sorting game.

The Attribute Game

This game is for as many people as you wish. Start with any one of the sets shown in Figure 42. The players study the characteristics or attributes of the figures and then make a decision. Discuss the selections the children make.

Figure 42. Sets for use in playing the attributes game.

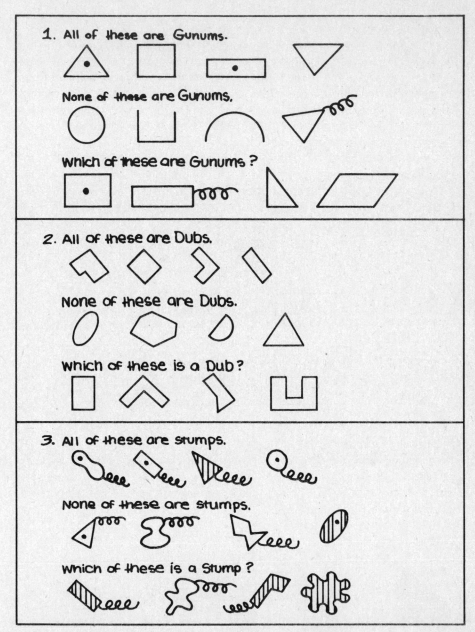

Inferring

Observations that we make are seldom enough to give us a complete picture of science. Going beyond or trying to explain observations makes up another very important science process called inferring. The first activity will give you an idea of what is involved in inferring.

The Footprint Puzzle[6]

Give a copy of the puzzle shown in Figure 43a to each child and ask, "What do you think happened here?" If some have trouble, ask a few questions such as:

1. What do you think made the tracks?
2. How many of them?
3. Which way were they going?
4. Were they running or walking?
5. Was the soil moist or dry?
6. Was the area hilly or flat?
7. What do you infer is happening where the footprints meet?

Carry on a discussion focusing on the inferences the children make and the observations they are based on. Tell the students you have a second piece of the puzzle that fits at the bottom. Tell them to be ready to revise some of their inferences (and yours, too). Turn the page and look at Figure 43b.

Figure 43a. Footprint puzzle.

Figure 43b. Footprint puzzle
(concluded).

Like your students, we suppose you want an explanation of the footprint puzzle. All we can tell you is that the prints are thought to be those of dinosaurs! No one knows what happened. But isn't that science. That is, it shows that science (inference) is tentative, and that rejecting old ideas and proposing new ones is part of the process. We hope this is an idea that is encouraged in your classroom.

We conclude then that much of science involves outcomes that are tentative, and subject to change and revision. The process of inferring should teach children this important idea. The next activities are designed to encourage children to make inferences based on their observations, and to know when they have made an inference as opposed to an observation.

Moldy Bread
Ten days before you do this activity, expose several pieces of bread to the air for about three hours. Put the exposed pieces of the bread in several dishes and cover them with plastic wrap. When you are ready to do the activity, divide your students into small groups and give each group three dishes with the contents as shown in Figure 44. Ask each group to make a list of explanations (inferences) for the differences they observe in each of the dishes. After the children have produced their lists, ask one child how he would test any one of his inferences.

Figure 44. Moldy bread activity.

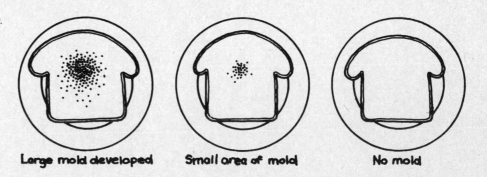

Large mold developed   Small area of mold   No mold

Figure 45. Where does the water go?

Where Does the Water Go?

Set up two pots of growing plants in your room with signs as shown in Figure 45. Then, for the next few days, have the class observe each plant and record their observations. The results will be something like those shown in Table 12.

Table 12.

		Plant Watering Observations	
Day	Time	Watered	Not Watered
1	8:30	looks good, healthy	looks good, healthy
2	8:30	looks okay, healthy	looks okay
3	8:30	appears healthy	slightly limp
4	8:30	healthy	losing color
5	8:30	healthy	bending over

When you think the children have made sufficient observations, ask them questions such as the following:

1. What inference can you make regarding the effect of water on plants?
2. Where did the water that you gave to the healthy plant go? (The answer to these questions is an inference, as illustrated in Figure 46.)

Figure 46. Where did the water you gave to the healthy plant go?

If you get responses such as these, ask other children if they think the statements are observations or inferences. All the student statements are inferences.

An alternative to this activity is for you to set up the experiment ahead of time and then present a healthy and a withered plant to the children. Focus can be on inferences based on the observations of the two plants.

Turning Process into Content

So far we have been "content free" with respect to concepts and content of science. (You haven't seen a section on electricity, or the human body, or machines have you? Well, don't expect it!) The "content" of science emerges from process, and you can make content exciting by emphasizing process.

How many times have you heard the teacher down the hall say, "This science book is dull; it has no science activities. It's one fact after another." Knowledge of process can turn that dull book into an exciting science program if you ask such questions as:

- "What things can I have the children observe that relate to the content in Chapter 1?"
- "Are there some pictures in Chapter 1 that I can cut out and use to get the children to make inferences?"
- "What things can the children bring in that will enable us to classify and measure?"

One technique we have already shown you is webbing (Chapter 1) in which activities can be generated from children's questions. Another way to turn content into process is to utilize the Process Generator Chart in Figure 47.

Figure 47. Process Generator Chart.

TEXTBOOK CONTENT		POSSIBLE PROCESS ACTIVITIES	TEXTBOOK CONTENT	
EARTH SCIENCE	LIFE SCIENCE		PHYSICAL SCIENCE	SPACE SCIENCE

rocks
minerals
fossils
sand
soil
road cuts
clouds
rivers
playground
rock shapes
weather stations
	plants
	leaves
	trees
	seeds
	insects
	fish
	birds
	animals under a microscope
	local woods
	pond
		Observing
		Have Your Kids
		Observe Things
			solids such as:
			   cloth, wood, metal ores
			liquids such as:
			   water, alcohol,
			   glycerine, detergents
			gases such as:
			   air, oxygen, hydrogen
			   helium
				moon
				stars
				sun
				pictures, photos of
				   planets, moon

rocks by texture minerals by color sand by size soil by color clouds by structure rivers by size fronts by shape storms by size earthquakes by depth	plants by height leaves by shape trees by bark color seeds by size insects by color fish by color	Classifying *Have Your Kids* *Classify Things*	sounds colors metals various shaped objects	planets galaxies
temperature pressure humidity wind tilt of rocks	plant height leaves tree circumference seeds insects fish	Measuring *Have Your Kids* *Measure Things*	mass length time areas volumes speed acceleration	positions of moon angle or height of stars motion of sun

Which of these 2 rocks
   do you think was
   formed in water?
Which layer of rock
   is oldest?
Why is this rock smooth?
	Why do raspberry bushes
	   have thorns?
	Why do seeds have a hard
	   surface?
	Why are the fish in the
	   aquarium brightly colored?
		Inferring
		Have Your Kids
		Make Inferences
			Why does the needle
			   float on water?
			How fast do different size
			   marbles drop in a jar
			   of water?
			Which candy bar gives you
			   the most candy for
			   your money?
				Why do you think the
				   moon rises 50
				   minutes later each
				   evening?
				Do the stars move
				   or does the
				   earth move?

It is designed to help you creatively turn textbook concepts and factual material into process activities. The chart is arranged so that the focus is on questions in which content is used as a vehicle for science process. For example, if you are working with your children on a unit on plants and seeds, your emphasis (from the chart) would be on:

1. observing seeds, plants, and leaves;
2. classifying varieties of plants and seeds;
3. measuring changes in plant growth; and
4. asking children how the amount of water or sunlight a plant receives affects its growth.

This is very different from emphasizing names of plants or identifying the parts of plants. Our suggestions are finite, but your ideas and those of your students are not.

Science Inquiry Problems

One of our goals as teachers should be to get children to think, to probe, and to inquire. So far in this chapter we have provided some ideas concerning sensory awareness and science processes. We see these as tools for inquiry, not as ends in themselves.

Usually we inquire into something that arouses our curiosity and interest. If some event doesn't fit with our past experiences, or we see a discrepancy, we usually ask why. We feel that setting up situations where children are presented with discrepant events or inquiry problems will foster student inquiry. More important, it activates interest, motivates children, and contributes to an exciting classroom.

Imagine the classroom scene pictured in Figure 48. Try the same activity with your class. Do you get similar responses? We hate to give you the answer in this book, but not doing so would make writing the next paragraph a bit difficult.

Figure 48.   The Great Pendulum Race.

The little person is right, both pendulums will swing back and forth at the same rate. Most will think the larger and heavier pendulum will swing faster. Now the discrepancy leads to questions and further probing. You might try something like what is shown in Figure 49. Give each child a chance to give his response. Then provide string and objects and tell each student to try his method for changing the motion of the pendulum.

Inquiry problems like these lead to student inquiry, questions, and provide an opportunity for student investigation. Here are some additional problems and activities that can lead to more student inquiry.

Figure 49. How can we change the motion of the pendulum?

The Clay Boat Inquiry[7]

Event

Show the children a ball of clay and a container of water. Drop the clay into the water (Figure 50). Challenge the children to make a piece of clay the same size float in water!

Figure 50. The Clay Boat Inquiry.

Guiding the Inquiry

Have available pieces of clay and containers of water. Do not help the children with a method for doing the inquiry. They will soon discover from ''messing around'' that they have to shape the clay into some form of a boat to make it float.

Further Inquiry

After the children have made boats that float, challenge each child to redesign his boat so that it will hold more materials (coins, paper clips, etc.) than anyones else's. This can be a boat-building contest if you provide the same amount of clay for each child.

- What types of boat designs are best for carrying large loads on rivers?
- What is the best design to get maximum speed in a sailboat?
- Who designs boats where you live? Where are they located?
- Can boats hold more than they weigh?

Moon Crater Inquiry

Event

Show the children two objects, both the same size but different shapes. A small pebble and a marble will do. Attach each to a piece of string about 50 centimeters in length. Ask them what the shape of the holes will be when you drop them into the tray of fine sand (flour is even better). Drop both objects from the same height. Have children observe both "craters" or "holes" (Figure 51). Why are they both the same?

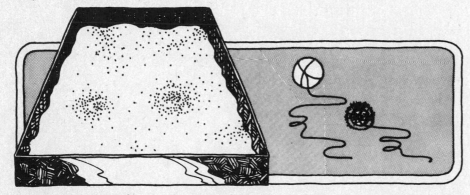

Figure 51.   Moon Crater Inquiry.

Guiding the Inquiry

Most of your students will be surprised by this discrepant event. You should encourage them to offer explanations or hypotheses to explain their observations.

This activity can be a motivator for the children to study the factors related to crater formation. Questions are posed in the further inquiry section which you may wish to use for discussion.

Further Inquiry

Provide fine sand or flour, flat pans, various objects to drop, string, and meter sticks for inquiry into the following questions.

- What effect does height have on the size and shape of the crater?
- What effect does the size of the falling object have on the crater's size and shape?
- What effect does the angle of impact have on the crater that is formed?
- Why does the moon have more craters on its surface than the earth?
- Are there other explanations besides the impact theory to explain the formation of craters?

The Attraction, Then Repulsion Inquiry*

Event

Tell the children to watch this event very carefully, that you are not going to say anything and that for them to find out anything they will have to ask questions that get a yes or no response from you.[8]

1. *Rub a comb or rubber rod with wool and bring it near a suspended pith ball (the inside of corn stalk, puffed wheat, or rice will do). The ball will be attracted (Figure 52a). Do this several times.*
2. *Now touch the comb to the ball. It is repelled (Figure 52b)*

Begin the inquiry by asking for questions.

Figure 52. Attraction, Then Repulsion Inquiry.

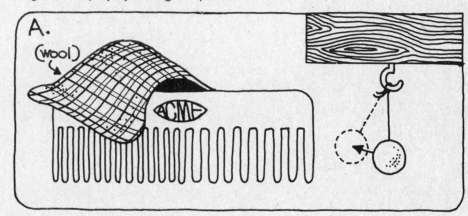

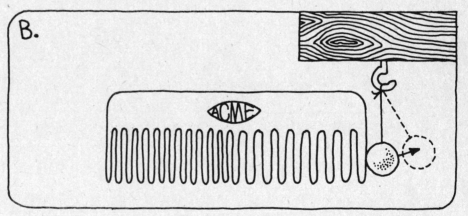

Guiding the Inquiry

Your role is important in this type of inquiry. Essentially you are telling the students that the only way they will get information about the event is if they ask questions like:

Question: Did you do anything to the comb after you rubbed it the first time?
Response: No
Question: Is that a plastic comb?
Response: Yes

When individual students have gathered sufficient information, you should encourage them to make declarative statements. These are also called hypotheses.

*This activity will work best on a nice, dry day.

Further Inquiry

Students can experiment with materials to test some of their hypotheses. If a student said, ''Like charges repel and unlike charges attract,'' then he should test other materials to check on the validity of his hypotheses.

Give the students other materials, such as strips of newspaper dropped over rulers to form leaves, silk and glass rods for rubbing, other spheres for pith balls.

The Muddy Current Inquiry

Event·

For this event prepare two slurries: (1) ½ baby food jar of water and ½ jar fine soil; (2) one baby food jar full of cold colored water. Nearly fill two clear tubes with water. Ask the children which of the two slurries will slide down the tube faster. Get as much involvement at this point as you can.

Pour both slurries into separate tubes at the same time (Figure 53). Have the children focus on which slurry moves faster. Ask questions such as: Was your original guess (prediction) correct? Why not? Did the mud speed up or slow down as it moved down the tube?

Figure 53. Muddy Current Inquiry.

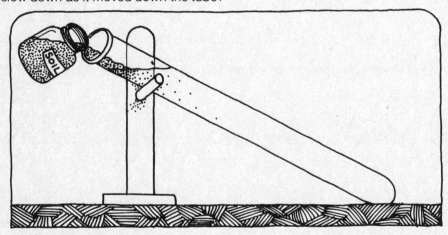

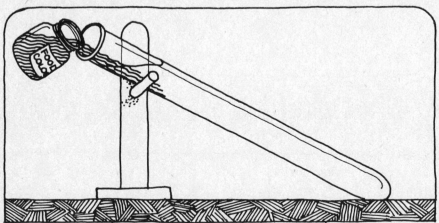

Guiding the Inquiry

This is a simulation of turbidity or density currents that occur near the edges of the continents, and within lakes and rivers. Let the children discuss some of their reasons for their predictions.

For Further Inquiry
- What will happen when a saltwater slurry is poured into a fresh water tube?
- How long does it take different density slurries to travel the tube?
- Record the position of the slurry every 2 seconds as it slides down the tube. Graph the results.

The Pendulum Inquiry[9]

Event
Can you guess which pendulum in each pair shown in Figure 54 will have the longer period? Try an experiment to find out.

Figure 54. Pendulum pairs.

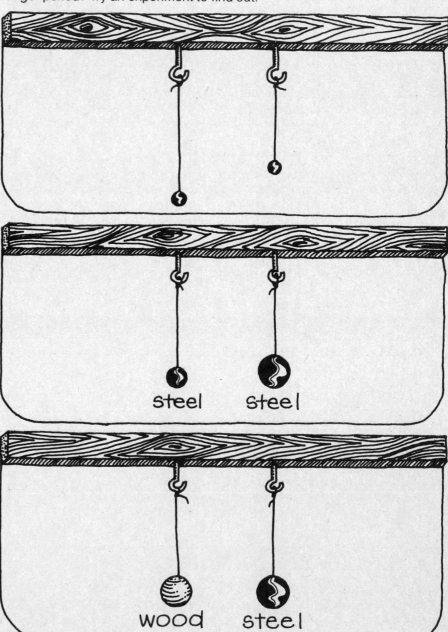

Guiding the Inquiry
Set this up somewhere in your room. Put the question on a card and allow the children to experiment when they wish. This is a good model for inquiries set up in a learning center.

For Further Inquiry
- How long must a pendulum be to have a 1 second period?
- A pendulum 1 meter long has a period of 2.007 seconds in New York City. At the North Pole its period is 2.004 seconds, at the equator it is 2.009 seconds, and in a balloon 98,000 feet above
- New York, it has a period of 2.016 seconds. Why are there such differences?

Conversation ═══════════════════

Joe: In looking back over this chapter I'm glad to note that the activities we've included really require very few resources.

Jack: Well, the process of discovery really just requires a child to use his natural resources, his senses.

Joe: And a teacher to develop an environment that includes plenty of activities that place the child directly in touch with his world.

Jack: Joe, what about teachers who feel that discovery will just happen naturally. In other words they would suggest that, when a child is left alone, he will spontaneously make discoveries, that having the teacher provide activities really restricts the possible direction for growth.

Joe: I'm sure that some might say that, but I think I would really disagree. I don't think it's very humanistic to disregard the training, knowledge, and experiences of the teacher and simply turn the child loose in a laissez faire setting.

Jack: In other words, the teacher's responsibility to the child is to provide enough quality activities so that the possibility of the child's really making discoveries is increased.

Joe: Well, I think that's pretty much a thread we've tried to keep throughout the book. The teacher is a person whose special training enables him or her to make some judgments about learning environments and select those components that will have the best chance to help.

Jack·   Then we really are saying that some experiences are more likely to help children make discoveries than others.

Joe: Exactly.

Cut and pasted construction paper.

8
Valuing

I saw the trees, the birds
The warmness shattered the protection of clothing.
I knew the sun, the trees, the birds, the bees, the ants, the grass, the
soil beneath my feet.
The crunch of grass and leaves echoed through the silent air.
A craving to stay, to run, to shout crept over me.
I restrained it.
The tension grew inside me.
I restrained it.
I burst!
I screamed.
I felt good.
I ran.
I leaped into a world unknown to me in reality, but always hidden inside
    me.
I felt happy, glad and good.
But now I climb back into the world of fault, of regularity, of reality.

———————————————————————————

I found a ladybug outside.
I let it crawl on me.
It tickled me.
But I couldn't keep it.
I wish I could.

These two poems written by fifth graders describe their immersion in another world, a world of feeling, wishes, and fantasy.[1] We wonder how many children's lives drift into a "world of fault, of regularity, of reality." In this chapter we hope to concern you with a dimension of learning, largely ignored by us as teachers (and by our culture for that matter), which you may perceive to be as important as cognitive learning. Hal Lyon states it this way:

> Learning can be enjoyable if it is humanized. What's more, learning which retains the human element is much more relevant to life. The intellectual must be coupled with the emotional if behavior is to retain a human quality. . . . The fact is, the intellect divorced from feelings is empty and meaningless. An education that is to be effective in preparing a child for life must take into account emotional as well as mental development. Schools must recognize that pleasure, and feelings are as vital, if not more vital, than intellectual achievement.[2]

We see valuing as the "other side" of the cognitive processes of discovering and exploring. In Chapter 6 we introduced the idea that some learnings could be characterized as logical and rational and termed them right-handed. Conversely, intuitive and aesthetic learnings were labeled left-handed. This chapter is about the latter. We hope to provide you with an array of activities and approaches that will move you from the right-handed or cognitive realm of learning and into the left-handed or affective arena. It is our hope that you will be willing to take this risk and make your classroom encounters more humanistic. We believe that children should have opportunities to express their feelings and emotions about what they are learning and about their world.

From Right to Left Learning in the Classroom

In Chapter 1 we alluded to the idea that a tremendous avoidance of science emerges in children as they move from elementary to secondary school. We think it has something to do with the notion that science has been taught one-handedly. (If some of you are thinking that the science courses you had were underhanded, we are not prepared to deal with that one here!) Science teaching that includes left- as well as right-handed learnings is something we need to consider.

Imagine you are teaching a group of fifth graders and you are beginning a unit on environmental science. The first activity you choose is one that involves the children in an outdoor science scavenger hunt. It primarily is a discovering or cognitive activity. You are hoping to bring the children into contact with the school environment via the scavenger hunt. The students, in pairs, are given the following list, a plastic bag, and a small vial. They are told to find and bring back objects in the school environment for each item on the list.

Scavenger Hunt[3]

1. Simple shapes in nature.
2. A simple machine.
3. A pleasant smell, an unpleasant smell.
4. Something that tastes sweet, something that tastes bitter.
5. A sound in nature.
6. Something older and something younger than you.
7. Three primary colors, three secondary colors.
8. Two or three different natural textures.
9. An animal trace.
10. A manufactured material being weathered.
11. A decomposer, a producer, a consumer.
12. A live animal to fit in a pill bottle.

When the students return, they sit in a circle to show and demonstrate the objects they have brought back. Naturally there is much discussion focused on the variety of objects they found. Although the scavenger hunt is fun, it is basically a cognitive activity. Each item on the list had a purpose related to the children's basic senses. To integrate the left-handed learning (affective) with right-handed learning (cognitive), you can propose three additional activities.

The first is a "Creepy Crawly Race." Using the live animal (most students bring back ants, spiders, a few worms, and some bees) and a "creepy crawly racetrack," the students engage in a live animal race (Figure 55). The first animal to the outside of the racetrack is the winner. The excitement and competition are fascinating. In fact, we've done this activity with teachers, and we see little difference in the excitement levels.

The second and third activities proposed are designed to provide the student with an opportunity to express his feelings about what he did during the scavenger hunt. The students are invited either to write a *syntu* or make an art form using a slice of wood and materials collected during the hunt.

Figure 55. Creepy Crawly Race.

For example, look at these poems written by some sixth graders.

Trees
Lovely, free
Blows, dances, rocks
Happily housing a bird
Logs.

Squirrels
Frisky, soft
Jumping, climbing, hunting
Hunting acorns and nuts
Scavengers.

Grass
Bright, green
Sprouting, jumping, poking
Working up through the ground.
Blades.

Wind
Cool, refreshing
Blowing, breezing by
Coolness in my face
Air.

Each of these poems is a *syntu*. The *syntu* is a Japanese poem consisting of five lines as follows:

Line 1—One-word name of an object, event, or phenomenon.
Line 2—An observation of the object using one of the five senses.
Line 3—A feeling about the object in line 1.
Line 4—Another observation of line 1 using one of the senses not used
　　　in line 2.
Line 5—A one-word synonym for line 1.

A session follows in which the students share their *syntu*s and art pieces with others. Somehow we lose sight of the importance of sharing our feelings with others, not to mention the fact that we have little opportunity to think about our feelings. As often as you can, you should provide time for your students to share their experiences with each other. And one last comment before we leave this scene. If you do this activity with your students, go on the scavenger hunt and write a *syntu* yourself, or do some environmental art.

Valuing—What's It All About?

Before we launch into presenting an array of valuing activities you could use in your science classroom, let's discuss what we mean by valuing. When we talk about valuing or left-handed activites in this book, we are referring to activities that focus on any one of the following: emotions, feelings, attitudes, beliefs, in short, the affective domain. Additionally, any of the following types of learning activities that you do would be considered within this domain: dreaming, fantasy, intuitive thinking, drawing, role playing, singing, and inventing.

We are interested in promoting the idea that science is a human endeavor, that it is not a sterile body of facts but an active, exciting process. Science becomes humanistic to the extent that we include the human side of learning. As with the example just described, the human side is obvious when a provision is made for students to express their feelings.

Carl Rogers has said that one of the most important conditions that facilitate learning is the quality of the interpersonal relationship between teacher and student, or to be more personal, between you and your students, and us and our students. Using valuing activities is probably the most effective way of developing and strengthening this bond. In dealing with our own students, we have found that our relationship is dependent upon: (1) how real we are, and (2) our awareness and understanding of our students' feelings. Trust takes time, as do realness and understanding. The process of becoming real and genuine, loving and caring, sympathetic and understanding is both joyous and painful.

Valuing Activities

We have organized the activities in this chapter into three categories: Warmups, Awareness Activities, and Value Lessons. *Warm-ups* are designed to help you and your students loosen up and get in tune with whatever situation you are in, for example, the classroom, at a creek (in a creek or up the creek!), in the woods. *Awareness Activities* focus on the psychomotor process of touching, hearing, seeing, tasting, and smelling to bring you and your students in contact with reality. Through these activities your students will be able to become aware of their feelings and attitudes toward the natural world of science. *Value Lessons* take you and the students deeper into the affective domain. You will be engaged in activities that involve action and decision making on issues related to science and people.

Warm-ups

We believe warm-ups are vital to the learning environment if you want each person to involve himself willingly. Many of the warm-up activities are nonverbal, which allows each person to immerse himself in activity without interference or threat from others. These brief, introductory activities are designed to get your students close to one another, loosen them up, and provide readiness for further activities. Many of these activities can be coupled with discovering activities presented in Chapter 7. By doing so you allow the student an opportunity to "get into" a playful learning mode.

Know Your Rock

Take the children outside and let each find a rock. Have them sit in a circle and tell them to touch and look at the rock so that they know their rock. Allow them a minute or so to observe their rock carefully. Have each person pass his rock to the left. Do this about four times, then have students change positions in the circle in a random fashion. Continue passing the rocks to the left, but now tell the students not to look at the rock as they receive one. Students are to distinguish their rock from the others, and as they do, they should remove themselves from the circle.

Naturally you can use other objects such as leaves, sticks, fruits, textured materials such as cloth, and so forth. This is a good powers-of-observation game, plus it's a lot of fun!

Hugging a Tree

Outdoors have each student select a tree. Tell them to get to know the tree by touching, feeling, smelling, looking, and even hugging the tree. If you do other activities outside, the student can focus on caring for his tree. You can extend this activity over a longer period of time. The student then can observe changes that occur to his tree over a longer period of time.

Blind Walk

In this activity the student will lose one of his senses (sight) and be forced to depend on his other senses to observe his environment. In pairs, with one student blindfolded (or with his eyes closed), lead the students around (outdoors is best, although you can do it in the school building). Try to walk over a variety of surfaces, up and down hills, under trees, and so forth. After 10 minutes or so, have the students reverse roles for a second walk. When the experience is over, conduct a discussion with the students by asking some questions such as:

1. How did it feel when you were blindfolded?
2. When did you start feeling comfortable, if at all?
3. What sorts of things did you observe when you were blindfolded?
4. What do you think of your partner?

Lying Down

This should be done outside. Have your students sit in a circle and join feet like the spokes of a wheel. Tell them that they are going to lie on their backs and observe what they can see, feel, hear, and smell. Tell them at times to close their eyes, and not to talk (if that's possible). After about 5 minutes, have the children sit up. Ask them a series of questions such as:

- How many animals did you see?
- How many kinds of sounds did you hear?
- What is the ground like around you?
- Do you like doing this?

Object Fantasy

Have the students close their eyes. Give them each an object (such as a twig) and ask them to explore it. Ask them to feel it with their fingertips, place it on their face, feel its weight. Have them imagine the twig is getting very large and that they are getting very small and that they are climbing around on it. Now have them return to the room very slowly and let those who wish share their experiences. You can use a variety of materials, such as pieces of science equipment, fruits, leaves, and balloons.

Fantasy Trips

Have the students lie on the floor and close their eyes. Lead them on a fantasy trip into a cave, desert, a mountain, or any place that seems appropriate for your students. Begin with something like the following:

> *Imagine you are floating on a river. The river is winding through a beautiful forest. You can see the trees, the beautiful golden flowers, the birds, the blue sky. Now the river reaches a mountain and flows into a cave. You float into the cave. Continue your journey and see what happens.*

Let about five minutes pass.

> *Now very slowly leave the cave and return to this room and to this group. When you feel like it, slowly open your eyes and sit up.* [4]

Bring the students into a circle and let those who wish to share their fantasy trips.

Awareness Activities

We have chosen to include a variety of awareness activities that emphasize a diversity of ways to help students become aware of themselves and their environment. Science begins with us and that with which we come in contact. These activities require that you risk, listen, feel, trust, be personal, and that you not judge what is right or wrong, good or bad, and so forth. Some of the activities (the first seven) have been modified from the Environmental Studies Project. [5] In these activities we are calling the assignment or activity an *excursion*. In most cases the student will have to go outside by himself or in a small group to explore part of his environment.

We have also included other types of activities based on the original excursion. You can use these activities in a variety of ways. You could use any one of the excursions with your entire class. If you do so, give the students the excursion assignment and then let them be free to decide what types of materials and methods they should use to complete the activity.

You could also provide options for your students by selecting three or four different excursions and letting groups of students decide which they would like to carry out. In any event, we urge you to do the activity with your students.

When the students return from their excursion, sharing the results should be optional; and if they choose to do so, make it informal. Sometimes simply displaying what they have done is sufficient. Other times some students will not be content until they are able to verbalize to others what they did.

Feelings: Good and Bad

Excursion: Go outside and find evidence for a good change and a bad change and a change that is neither good nor bad.

For Further Action

- Discuss with students what good and bad mean to them.
- Encourage the children to engage in good changes occurring in their community.
- Do an art activity creating signs depicting good and bad changes.

Power

Excursion: Go outside and take pictures (or draw pictures) that represent power. Exchange pictures or drawings with other students and discuss what kind of power you see in each person's pictures.

For Further Action
- What is power?
- Where does power come from?
- Write a *syntu* about power.
- Do you have power?

Touching

Excursion: In small groups, and blindfolded, pass around a variety of objects (use such things as bags of sand, rocks, fruits, meat, various containers, balloons filled with water, etc.). Communicate your feelings to each other.

For Further Action
- Discuss what you learned about yourself from the experience.
- Do you like to touch things?
- What can you find out about things by just using the sense of touch?

Loving/Hating

Excursion: Go outside and bring back objects you love and things you hate.

For Further Action
- How do your love things compare to other students' love things? Hate things?
- Using your objects make an art form such as a poster or a collage.
- Which of your objects do you love the most? Hate the most? How do other students' selections compare?

Sounds of Nature

Excursion: With a tape recorder go outside your classroom and find and record sounds that you like and dislike; sounds man-made and natural; morning, afternoon, and night sounds.

For Further Action
- Did you find any sounds you could bring back without a tape recorder?
- How do sounds make you feel?
- Are some sounds pollutants?
- Write a *syntu* about a sound.
- Use some pictures or slides and match sounds you found with pictures.
- Show them!

Art and the Environment

Excursion: Find and collect materials in your local environment and make some form of art from them.

For Further Action
- Look at other students' art forms and find out:
  - —how ugly the environment is
  - —how beautiful the environment is
  - —the sadness of it
  - —how they make you feel
- Go back to the environments and find places that you want to improve.

Opposite Objects

Excursion: Make a list of opposite word pairs, such as happy–sad, friendly–hostile, afraid–secure, and then find objects in your environment that can be represented by the word pairs.

For Further Action
- Set up posters displaying your word pairs and objects.
- Return to the environment and find objects that present the same idea or attitude.
- Classify objects you find into categories of attitudes.

Seeing with a Camera

Excursion: Go outside in the school ground area and identify objects, features, or phenomena that make you feel: happy, sad, angry, beautiful, hostile, afraid, disgusted, massive. Take photographs of the objects. Keep a record of your "feelings" and the pictures on an index card.

For Further Action
- Find out how others feel about the environment. Show them your pictures and a list of the feelings and see if they can match feelings with pictures.
- Do others feel the same way you do about the environment?
- What did you learn about the environment?

Body Mapping

Using a large roll of paper (butcher paper), cut lengths long enough for one student to lie on. Have another student trace the outline of the student's body with a crayon. Each student should have his body outlined. Then in pairs they should "map" their body by gluing pictures from magazines and newspapers that tell about themselves on various parts of their body map.

Science Value Poster

Children love an opportunity to work with art materials in science. In this excursion they have a chance to use art and science as they express their values. Give each child a sheet of unlined paper, oak tag, or construction paper. Have the children use a crayon or marking pen to divide the surface as shown in Figure 56. Then encourage them to use their artistic talent to *draw* their answer to the following six questions, placing the drawing in the box whose number corresponds to that of the question.

1. What kind of animal do you think you most like?
2. What is your favorite activity?
3. What do you do best?
4. What is something that you are striving to become?

5. What about science do you like most?
6. What is one thing you would like other people to say about you?

When the children have finished, have them sit in small groups and tell about each of their six drawings.

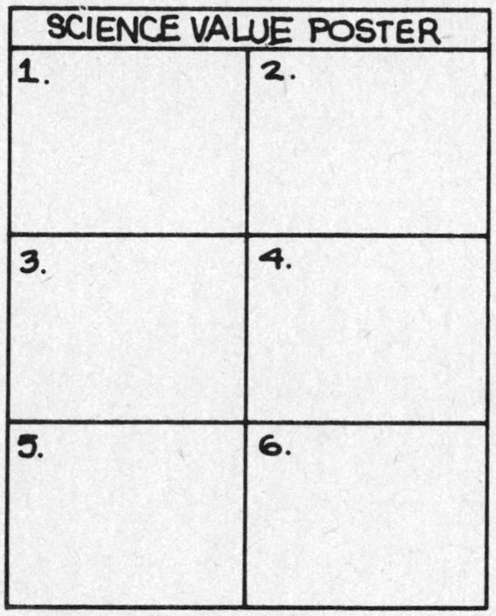

Figure 56.

Music, Poetry, and Science

Find some books that have a variety of poems and select some that are science oriented. Poems can capture the beauty, tragedy, joy, sorrow, and excitement of the natural world. They are good vehicles for students to deal with their feelings.

Select and play some music you think your students will like while you read the poetry to them. Here is a sample poem and how you might use it. (The sounds of the ocean would be appropriate here.)

After the students have listened to the poem, ask them to close their eyes, put their hands over their ears, and imagine that they live inside a shell. Tell them to shut out all sense of themselves and really try to be a shell creature. Help them by asking them to act out (in their mind) the following:

· How do you feel inside your shell?
  (Do you feel cold or warm? Alone or protected? Relaxed or cramped?)
· How do you move?
· What do you see?
· What sounds do you hear?
· What texture is your body?

Ask the children to open their eyes and then ask if any would care to share their experience. After this activity, discuss with your students the possibility of having a poetry reading in the near future. Suggest they bring one poem to read and a favorite record to play.

The Matching Embryos Game[7]

Give each student a copy of the sketches shown in Figure 57. Then give them the following information:

SHELL SONG

*Under my shell
My pearly shell
My steely shell
My crystal shell
Alone and alone
Alone I dwell.*

*Huddled in a head
My low bent head
My unbowed head
In its undried blood
Listen at sound
That gives no word.*

*I call from within
My tunneled wall
Hearing the echo
Of my call,
Sure in the color
Of my shell.*

*More sure of alone
And alone and still.* [6]

Figure 57. Embryos for matching.

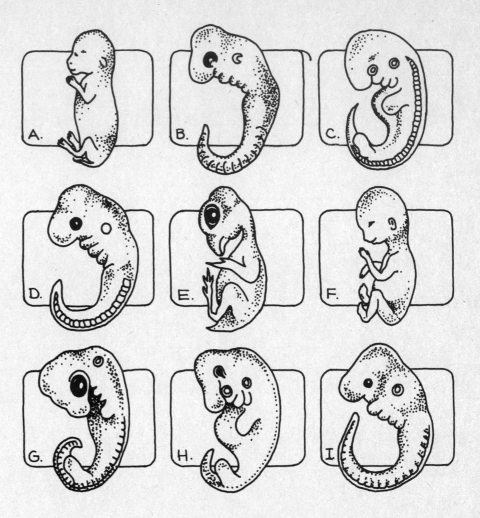

Pictured on this page are sketches of embryos. The sketches are coded A through I. Three of them are of a chick, three of a rabbit, and three of a human. The three of each creature represent successive stages of growth of the embryo of that creature. Use your perceptiveness (and luck) to determine which three sketches represent the embryo of each creature and then decide which of each three represent the first, second, and third stages of growth for that creature. Record your decisions (rabbit 1, rabbit 2, etc.) on the line at the bottom of each sketch.

Discuss the answers (A-rabbit 3, B-chick 1, C-rabbit 2, D-human 1, E-chick 3, F-human 3, G-chick 2, H-human 2, I-rabbit 1) with the students. Then ask questions such as:

- How does this make you feel?
- How are we different from other animals?
- Did you realize there was so much similarity among animal embryos?

Value Lessons

Value lessons are activities designed to involve you and your students in issue-related aspects of science teaching. They are designed to elicit opinions, enable students to take positions, and indicate what their beliefs are. Very few issues can be handled in a simplistic, yes–no, right–wrong, black–white manner. In these activities students should be given the opportunities to explore their feelings and those of others in a way that is nonthreatening. The emphasis should be on learning, relearning, and discovering what their own values are.

Science Soapbox

The purpose of this activity is to have your children express their opinions and values about topics related to science and your class. Have the children sit in a circle either at their desks or on the floor. Use the questions we have provided as a starting point for class discussions. We think these class discussions will help you and your students get to know each other and provide a forum for everyone to express his or her feelings. For any given class discussion you may wish to discuss only a few topics. Ask one question at a time and allow as much time for discussion as there appears to be interest.

Sample Questions

1. What kind of scenery do you like best?
2. If you could take a trip anywhere, where would you go?
3. Will you become a cigarette smoker? Why?
4. Do you like science fiction movies or books?
5. What's your favorite breakfast?
6. Are you a good eater?
7. Do you think it's easier being a boy or girl?
8. What's your favorite food?
9. What are your favorite things when you go outdoors?
10. Would you like to be a weatherman (geologist, biologist, chemist)?
11. Would you like to become a great cook?
12. When you grow up, what kind of job would you like to have?
13. Do you think people who take dope are smart?
14. What's your favorite kind of pet?
15. Which do you like better, the mountains or flat land?
16. When people die, what do you think happens to them?
17. Do you believe in witchcraft?
18. If you were to travel to Mars, do you think you'd find life there?
19. Are there such things as flying saucers?
20. What do you think a scientist is like?
21. Do you know of any polluters in your neighborhood?
22. Do you usually feel healthy?
23. Would you like to be a scientist (science teacher) some day?
24. What do you like least about this class?
25. What do you like best about this class?
26. Do you believe animals and plants developed through evolution?
27. Do you believe in love?
28. Do you think nuclear power has been beneficial to us?
29. What do you think should be done to people who do not take care of their pets?
30. Do you think we should keep animals in zoos?
31. Do you have full polio protection?
32. Do you wear seat belts?
33. Do you like to meet children who are different from you?

34. Have you ever invented anything? What?
35. Do you prefer warm or cool days?
36. What do you do on rainy days?
37. Do you like boys (girls)?
38. Do you like to go shopping with your parents?
39. Do you do any gardening?
40. Do you have any house plants?
41. Do you think science has all the answers to our problems?
42. Do you think working in a laboratory would be fun? Why?
43. If someone is definitely dying, do you think we should keep him alive with medicines?
44. Do you think scientific laws can be changed?
45. In what ways is science important to you?
46. Do you enjoy studying science?
47. Do you think scientists agree with each other?
48. How do you feel in a thunderstorm?
49. What's your favorite topic in science?
50. What do you think about the cleanliness of the water we drink and the air we breathe?

Project Work Using Sticky Science Issues

What do students in your class think about various issues that are related not only to science but to their personal lives? This value lesson is designed to have the students publicly indicate their opinions on a variety of issues. We have included a number of issues here, but we suggest that you solicit issues from your own students (give each student a 3 x 5 inch index card and have him write one issue he would like to have considered).

You can use the issues listed here in a variety of ways with your students. Two ways we suggest are an opinion poll and a debate.

Opinion Poll.   Give the issue list or a portion of it to groups of about four students. Suggest that they select about five related statements from it to use in a public opinion questionnaire. They can then solicit opinions from other students in the school, adults in the community, and their parents. If they wish, they can share the poll results with the class.

Debate.   Select issues on which you think there will be differences of opinion. Suggest that two or three students form a group to "research" the issue and prepare to defend their position in a debate.

Issue List

1. Scientists are very brilliant people.
2. Pesticides should not be used near people's homes.
3. We are responsible to and for one another.
4. The oil crisis was the result of oil companies wanting higher prices.
5. Smoking should not be allowed in schools.
6. If a student doesn't learn, it's his or her own fault.
7. Animals should be used for science research even if they are killed.
8. Drinking liquor is bad for you.
9. A law should be passed making people use seat belts.
10. The government should have the right to put people who pollute the environment in jail.

11. Gasoline should be rationed.
12. Car pools should be required so we can conserve energy.
13. Man should continue to explore space even if it is expensive.
14. Man is taking better care of his environment.
15. Dogcatchers should be outlawed.
16. Instead of arguing about ideas, everyone should agree to give ideas to others.
17. Anyone can be an artist.
18. If one scientist says an idea is true, most other scientists will agree with him.
19. It's okay to daydream in science class.
20. Would you go to the moon if you had a chance?
21. Most people believe in magic.
22. Astrology is a science.
23. Taking walks is fun.
24. Drug pushers should be put in jail forever.
25. Parents should discuss sex with their children.
26. Do you want to be a scientist someday?
27. People in general are good to their pets.
28. Playing with dolls is okay for boys.
29. Parents generally talk about childbirth with their children.
30. Science is a collection of facts.
31. Science is too hard for me.
32. Science is important in my life.
33. I like science.
34. Someday I would like to be a scientist.
35. Weathermen are usually correct when they make their forecasts.

Open Forum[8]

Many of the issues you and your students will discuss in the science classroom cannot be resolved in black or white, yes or no, agree or disagree terms. Your students may view each issue from several positions.

Divide your class into groups of five to eight students. Each group chooses or is assigned a controversial issue. The ones listed below are samples; the ones you make up may be more meaningful to your class.

mercy killing
zero population growth
nuclear power development
use of pesticides
women in science
exploring space
legalization of marijuana
humane use of animals
responsibility for care of environment
world food problem
use of automobiles in large cities

Each group identifies as many possible positions on the issue as there are people in the group. This identification process will be an open forum where each person presents his position. On a large sheet of paper have the students in each group arrange their positions along a line, a continuum, with the extreme positions on the ends. When all members of all groups have defined their relative positions, bring them back together and have them share their positions with the entire class.

Value Continua[9]

In this value lesson students will be asked to indicate their position on an issue by placing themselves on a continuum for a number of science-related issues. By defining the ends of each issue with a brief description or representative names for the issue, the student can take a position relative to the extremes. The following are some issues and extremist people.[10]

Drinking	Boozer Barry	Sober Sam
Racism	Red Neck LeRoy	Liberal Larry
Birth control	Bulging Bertha	Pill Poppin Patty
Maturity	Baby Bobby	Mature Mike
Smoking cigarettes	Lungless Larry	Smokeless Sam
Personality	Wallflower Willie	Outgoing Otis
Buying	Hard Cash Harry	Credit Card Cathy
Quantity of food eaten	Gary Glutton	Skinny Mini
Clothes	Sloppy Joe	Fashionable Theo
Religion	Andy Atheist	Holy Holly
Personal hygiene	Sally Stench	Mouthwash Mary
Quality of food eaten	Hot Dog Hank	Health Food Harry
Sports	Peter Pansy	Joe Jock

Other issues could be taking care of animals or plants, space travel, use of drugs, taking care of children, and so forth.

You can do this activity several ways. Draw a line across the chalkboard or on a large piece of paper and mark the ends of the line with the issue. For example, if you choose personal hygiene, Sally Stench would be written on the left end of the line, and Mouthwash Mary on the right. Have the children come to the board and put their initials somewhere on the line representing their position. You should do this too!

Another way is to distribute value continuum worksheets (Figure 58) and let each student mark his own sheet.

Figure 58. Value continuum worksheet.

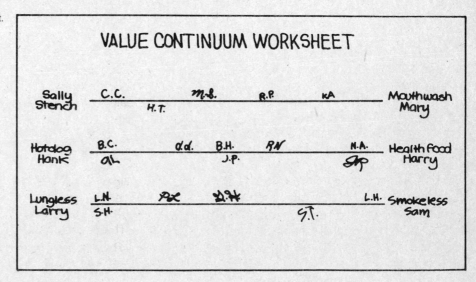

Role Playing[11]

Role playing is a process of acting out situations, events, people, and processes of which we are not really a part. We have suggested several role playing situations that you could use in a variety of classroom settings. If you are interested in spontaneity and creativity, then this value lesson is one you'll want to try.

You should plan to include three phases using any one of the scenes depicted below. The phases are warm-up, enactment, and sharing. The warm-up should establish the mood or set for the role playing enactment. Using whatever technique available (pictures, film, discussion, reminder of prior activities), it is important to bring the students to a point where they are intellectually and emotionally close to the content and process of the scene you are going to act out.

The enactment is where students act out the scene. For example, if you are enacting a landing on another planet and discovery of some form of life, you might have the students build props such as planetary landscape. The students should be encouraged to "make up" their own characters and act them out. Judgment and evaluation of the acting should not be permitted, but discussion of what is happening should be.

The sharing scene is crucial for the students so that they can share their personal experience in the scene. Those who make up the audience should be encouraged to give their perceptions and express their feelings.

The following are some examples of scenes you could create and act out in a role playing setting.

- Enact what it is like to be a red blood cell in the circulatory system of the body.
- Portray a trip to the center of the earth.
- Convince someone who is not aware that the earth rotates.
- Imagine yourself as one part of the human body and explain your structure and function.
- Enact a bird learning to fly.
- Enact a debate in congress about a clean air bill.
- Enact a court scene where some industry is being prosecuted for polluting the local air and a major river.
- Enact weightlessness as an astronaut does in space.
- Select a concept and have the children act it out using their bodies. Some possible concepts are: light, gravity, mass, volume, speed.

Conversation

Jack: Valuing activities really bring out a part of science education for children that is seldom considered.

Joe: Well, people do think of the school setting as a place to encourage the development of values, but only rarely do we see valuing in a science program.

Jack: I guess project work in the social studies area and children's literature for the language arts usually provides some valuing experiences, but that's about it.

Joe: It's really a shame because children need to participate in valuing within the science program. I think it really helps break down the stereotypes about scientists fostered by the mass media.

Jack: I guess you mean the strange little men with the white hair working alone up in a laboratory.

Joe: Right. By doing valuing activities we can encourage children to view "sciencing" as something done by humans.

Jack: The valuing activities sure do bring that out.

Joe: And besides, they're fun to do!

Cut and pasted construction paper.

9
Exploring

"The child is first" is the way we started out viewing the learning process earlier in the book. We still believe it! With exploring, the child is first. We view the child, an inquirer, as a little person who, aside from some stimulation, needs the freedom to grow and become fully human. The child is an explorer moving in all directions, being bombarded by the stimuli of his environment. To grow to become what he can become requires that he be given the encouragement and the freedom by us as teachers to explore and to pursue his own goals and interests. We envision the child exploring his world in many ways, from stopping on the way to school to turn over a rock and watch a bug to reading about insects in a book.

In the classroom we must provide many alternatives and options for the child to explore. Earlier in this book we discussed the need for a variety of individual, small group, and large group experiences. We are not all alike, and we know that we learn in different ways, ways that are unique to each of us. Our classrooms can encourage children to become explorers by providing activities that permit them to explore through:

1. student projects

2. field trips

3. student experiments

4. learning centers

We hope to provide you with some helpful ideas about these four approaches to exploring in this chapter.

Student Projects

Imagine watching one of your students bringing in from home cardboard, tools, nails, glue, paint, and three books on rockets from the library. You're delighted! She has not done very much during the year; but the other day when you announced that the students could work on a science project from one o'clock to three o'clock, three days a week, Martha suddenly got excited. All of a sudden, building a model of an Apollo space rocket has turned her on.

Projects can do just that. They can be done individually, in small groups, or by the entire class. The best projects are those the children decide to do, not the ones we tell them to do. They also will be better if they are not evaluated by you. A project represents a child's approach and method which we feel taps his or her creativity. When we impose standards and evaluation schemes, we inhibit creativity, and the project turns out to be something we want, not what the student wants.

If you decide to include projects as part of your science program, you may want to stock your room with a wide variety of odds and ends. Here are some examples: hammers, saws, wood, glue, paint, film, cameras, paper, cardboard, nails, tacks, bricks, sand, rocks, cloth rags, maps, etc. The most inexpensive way to get these items is by making up a scrounge list, such as the one in Chapter 3, and sending it out with a cover letter.

In this section we are going to suggest a number of projects that have proved to be of high interest with other teachers and students. We hope they are useful to you and to your students; however, helping your students develop their own ideas is even better.

The Egg Drop

If you are studying forces or motion, this project might be of interest. Students are asked to build a package in which one raw egg will withstand a drop of about 8 meters. The egg must survive! Also the size of the package should be restricted to about 20 centimeters in length, width, and height (see Figure 59).

Figure 59.   The Egg Drop.

An additional part of this project can be to develop some mechanism for lifting the egg package the 8 meters. How about a hot air balloon?

This is a great class project!

Rock Polishing

This is a long-term project. It takes about 30 days to polish minerals (gems) in a rock tumbler. One of your students may have a rock tumbler at home and may be willing to let you use it in your room.

Minerals with a hardness around 7 (quartz, agate) polish best. The geology department at a local university is a good source for free samples of minerals, and you can purchase samples at a rock shop. Of course, it is even better for you and your students to collect them.

If you have a few students interested in this project, take them to a rock shop for instructions on how to use the rock tumbler. When the rocks are polished, the students can make a variety of jewelry such as rings, bracelets, pendants, and earrings.

Building a Weather Station

Most of your students will probably be interested in exploring the weather. One of the best ways to do this is to have some of your students build a weather station. Encourage students who are interested to construct their own instruments. The only instruments you will need are shown in Figure 60.

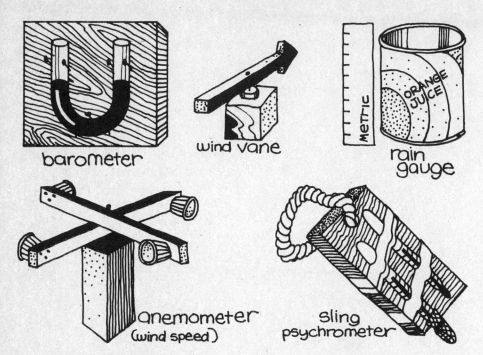

Figure 60. Student-made weather instruments.

barometer

wind vane

rain gauge

anemometer
(wind speed)

sling
psychrometer

Most intermediate school science textbooks have a section on weather study that includes all the information you need on how to construct weather instruments with simple materials. The students who make the instruments can begin collecting weather data and recording the information on a daily weather chart. They can keep track of such things as wind speed (number of turns of an anemometer per minute), wind direction, and percentage of cloud cover in the sky. You might even start off each day with a weather report and forecast from your junior weatherpersons!

weather station

Animal Tracks

Another interesting project, especially if some of your children live near a small creek or if the school grounds are located near such an area, is to make and collect casts of animal tracks. Footprints of a variety of animals, birds, squirrels, chipmunks, opossum, deer, dogs, cats, and so forth are abundant (you can literally have casts of thousands) in the mud or flood plain of a small stream or creek.

A simple collecting technique using plaster of paris and a piece of cardboard is illustrated in Figure 61. After the animal track casts have been made, further exploration will follow. The children will want to know such things as:

1. What animals made the tracks?
2. When were they made?
3. What were the animals doing?

This project can lead the children into a study of the types of animals that inhabit their area.

Figure 61. Techniques for collecting animal tracks.

1. Clean track and spray with shellac or plastic.

2. Pour plaster of Paris into circle of cardboard surrounding track.

3. When hard, lift cast. Then clean it carefully and coat with vaseline.

4. Pour more plaster over casting surrounded by wide strip - level w/top.

5. Separate two layers of casting. Clean and smooth with knife.

6. When cast is dry, paint inside of track with black India ink.

Added Attractions

1. Making photographs of the environment.
2. Making collections of rocks, minerals, fossils, shells.
3. Making collections of butterflies, insects.
4. Collecting pictures and making scrapbooks on various topics such as: big things, little things, round things, square things, colored things, ugly things, beautiful things, mountains, lakes, vacation places, where organisms live, etc.
5. Making a garden inside or outside the classroom which would include a variety of flowers and vegetables. (How about an organic garden?)
6. Making models of rockets, volcanoes, atoms, planets, cities, rivers, animals, people.

7.   Writing plays and stories about science and acting them out.
8.   Obtaining a soil test kit and having students analyze the soil in the
     school yard and at their homes.

Field Trips

Does this cartoon remind you of field trips you might have taken in science
class when you were a student? Unfortunately, this is a typical misconception
of field trips.

A second misconception about field trips we want to dispel is that you need
a bus, and you must drive 200 miles and listen to someone say, "This mountain
was formed 1 billion years ago and is made of granite" and then turn around
and drive home! And finally, do not plan field trips as show and tell activities.
Involve the students. And the best place to involve the children is from the very
beginning, in the selection of the field trip itself.

We believe field trips can be a rich source of learning and exploration for
you and your students; but because of school regulations and the problems
associated with leaving the classroom, most teachers tend not to include them
in their teaching plans. Maybe it is because we have very traditional views of
field trips. We are going to make suggestions for two types of field trips we think
will be attractive to you, and may eliminate some of the problems you may have
had in the past.

Mini Field Trip Ideas[1]

Mini field trips by their nature involve short, intensive contacts with the child's
environment. We recommend that you keep them to 10 to 20 minutes in length.
Also, because they are brief, you will be confined to exploring the environment
around your school. We recommend that you transfer the mini field trip ideas to
laminated 5 x 8 inch index cards so that you can use them in a variety of ways,
for example, as entire class, small group, individual classroom, or home assignments. Figure 62 illustrates how you can do this.

Figure 62. Mini field trip idea cards.

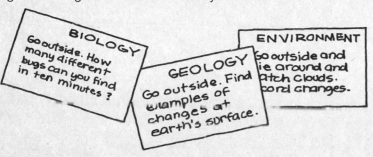

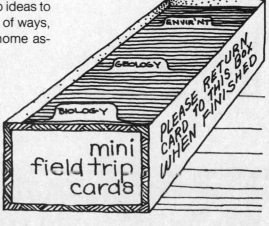

Appeteasers

1. Go outside and fly kites to find out about the wind; let helium and air-filled balloons go aloft.
2. Lie on your back outside and watch the clouds. Do this on a day when there are cumulus clouds. Watch them disappear, change shape, and form.
3. Give students a shoe box containing things like string, rulers, compass, hand lens, pencil, paper, thermometer, litmus paper, containers, and plastic bags. Tell them to find out as much as they can about a particular area in a limited time.

4. Make a map of the school grounds. Be as detailed as you can.
5. Go down to a creek or a pond and collect a baby food jar full of water. Examine the water under a microscope or magnifying glass. Make drawings of what you see.
6. Take a tape recorder and record natural and unnatural sounds.

7. Measure the outside air temperature once each hour, starting at 8:00 A.M., until the end of the school day. Graph the data.
8. Go outside and find as many examples as you can of changes taking place at the earth's surface.
9. Find as many living things as you can in an area in the woods 2 meters square.
10. Make a map of the 2 meter square area showing how you think one living thing affects another.

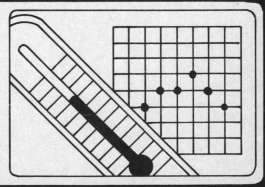

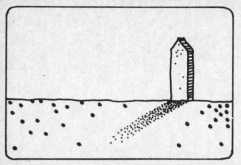

11. Stake out a small area outside and plant a garden. Plant a variety of things. Watch them grow. How fast do they grow? Which plants grow fastest?

12. Obtain a net and then go outside and collect several varieties of insects.

13. On a sunny day set up a stake in the playground. Study the shadow cast by the sun by marking its position every hour. Repeat at various times during the year.

14. Get up early some morning. Where does the sun rise? Make a sketch to show it rising in relation to familiar trees and buildings. Repeat again about a week later, a month later. Did it rise in the same place each time?

15. Stake out an area about 1 meter square. Analyze the temperature, moisture, color, and texture of the soil in the plot. Make a small map of the soil area.

16. Prepare a pile of dirt outside on the playground (about 1 meter in diameter). Return to the pile each day to make sketches or take pictures. How long does it take for the pile to erode?

17. Study the bricks on the outside walls of your school building. Are there any differences among the north, south, east, and west facing walls? What might cause the differences?

18. Visit a graveyard. Do grave markers weather at the same rate? Pick out one year, at least 50 years ago, and compare grave markers.

19. If there is a stream nearby, study how fast it is moving. Drop corks in and time how fast they take to go measured distances. Collect samples of the water with large baby food jars. How much and what kind of sediment does the stream carry?

20. Place various pieces of material outside such as fresh rock, aluminum, glass, tin, chrome, copper. How long does it take for each material to begin to weather?

Yellow Page Field Trips

That's right, let your fingers do the walking. The yellow pages of your phone book are the best references you have for places to visit that involve science and people. That's the purpose of this category of field trips. Too often we forget that a good percentage of people in our community work in occupations and businesses that are science related.

For instance, consider this list of categories that appears in most yellow pages of the phone book and the corresponding sample of the things we could learn about.

Category	Learnings
airport	principles of flight weather radar
animal hospital	pets medicine
carpenter	measuring structures wood products
chemical	how chemicals are made chemical products
florist	plants soil flowers
hypnotists	concentration the human mind psychology
photographic equipment	taking pictures chemistry lenses light
printers	machines
sewage disposal system	water filtering biology

Let's consider how to make use of the yellow pages as an invitation to explore the community in which you teach. Suppose you and a group of fifth graders are studying a unit on life science and the students wish to take a field trip. A sample of some of the places they could choose from appears in Figure 63. The list of people and places to visit will be lengthy for most topics you choose; thus you will have a selection problem which will have to be worked out with the children.

Once a decision is made, several possibilities are available with respect to the place you have chosen. Suppose it is a health food store. You could:
1. Take the entire class.
2. Have a parent take a smaller group.
3. Invite someone from the health food store to your classroom.

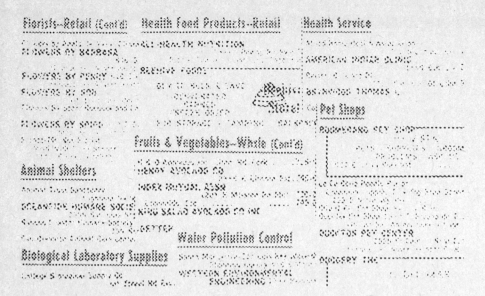

Figure 63. Life science field trips.[3]

Regardless of what you decide to do, the basic question your children will be concerned with is, ''What can we learn from a health food store?'' For example, here is a sample of questions children might ask:

1. How do you make granola?
2. With what material do you make granola?
3. How hot is the oven?
4. How much do the various cereals and dried fruits cost?
5. Is granola really more nutritious than Winkle, Tinkle, and Poop cereal?
6. Is it easy or hard to make granola?
7. How can I make granola at home?

Another way to use the yellow pages is to focus on science-related occupations. The new emphasis on career education can be brought into your science program through these field trips. Figure 64 is a group of yellow page selections that you could use in this way. What careers could you and your students explore at these places?

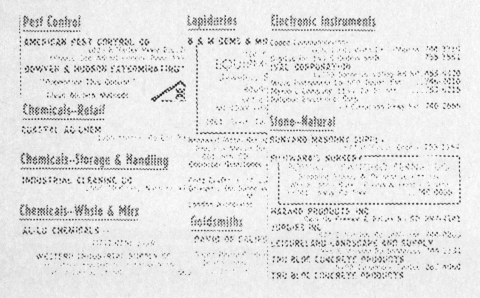

Figure 64. Career field trips.[4]

Student Experiments

Science to the layman is usually linked to experiments. How many times have you seen pictures of the "scientist" in *his* white coat peering at some bubbling test tubes and bottles? Yet, experimenting is more than manipulating chemicals and test tubes. It is an exploring process in which we have posed a question or two about some particular event, process, or phenomenon, controlled an array of conditions, and in most instances collected data to test the question we originally posed. And the questions we ask are not pulled out of a hat, but rather are based on some convictions we have about reality.

THE SCIENTIST
DOES NOT STUDY NATURE
BECAUSE IT IS USEFUL;
HE STUDIES IT
BECAUSE HE DELIGHTS IN IT,
AND HE DELIGHTS IN IT
BECAUSE IT IS BEAUTIFUL.
— H. POINCARE

Experimenting involves several substages or processes. We have chosen to reduce the list to the following:

1. model building
2. controlling variables
3. formulating hypotheses
4. designing experiments

Children will be motivated to engage in experimenting if they have learned some of the tools of this process. We suggest you involve your students in group activities that focus on these four processes. Individual exploration and experimentation will probably increase if children have confidence in their skills.

Model Building

Making sense of observations, searching for unifying patterns, and developing a mental picture of reality are all ways of expressing the meaning of model building. It is a process that is full of creativity and inventiveness. It is a process where we can break from tradition and propose new ways of looking at reality and pose new models. A model is a person's representation of how he perceives reality. Sometimes we get carried away with the personal involvement we have in our own model and as a result will not listen to others. Sometimes our models are rejected as being impossible (continental drift idea) or heresy (Galileo's concept of earth and sun motions).

The most important thing about a model is that it is a model! It's a conception of something, and it should be viewed as tentative, partial, and subject to change. Here are a few activities designed to help increase your students' model-building power.

Make It Light!

Give each of your students a battery, small light bulb, and a single piece of wire (Figure 65). Ask them to put the objects together in such a way that the bulb lights. Don't give them any help, but do give them plenty of time. When everyone has made the bulb light, ask them to show in a diagram or drawing how

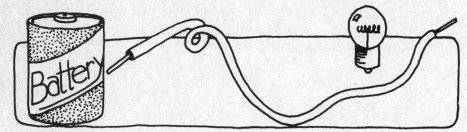

Figure 65.   Make it light!

they did it. Have them label this drawing Model # 1. Now ask them if they can make the light come on another way. When they can do it, have them make another drawing and label it Model # 2. Let them try a third way and produce a third model (Figure 66). Have the students discuss and compare their models. Which is the most accurate model? The best model? Questions like these will encourage the children to see there are alternative ways of viewing events.

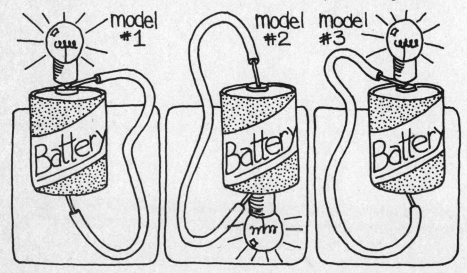

Figure 66.   Three alternative models.

Inquiry Boxes

An inquiry box is a closed shoe box. The interior of each of the boxes is differ-ent. A small marble or ping-pong ball within the box can be used to find out what the inside of the box is like. By tilting the box in various directions and by carefully listening to the sound, the child can make a diagram of his guess about the inside of the box. The diagram in Figure 67 shows how to make the inquiry boxes. You and the children can make many more.

Figure 67. Making an inquiry box.

Give groups of students a set of inquiry boxes and tell them to find out what they think the inside of the box looks like. Tell them they should make a diagram for each box. (Incidentally, only the persons who make the boxes know what the interiors look like!)

When all groups have finished, have each group show what their model of each box is by drawing it on large sheets of paper or on the chalkboard. Parti-tioning the chalkboard into as many sections as there are boxes and providing as many exterior outlines of the boxes as there are groups will facilitate com-munication and discussion.

The discussion can focus on such questions as:
1. How are the patterns for the boxes alike or different?
2. What do you think about other groups' models?
3. How could we improve the models that we have made?

Pictograms

This activity is designed to help you and your students become aware that people have very different conceptions, even though they perceive the same objects, collection of objects, or phenomena. We have provided two illustra-tions (Figures 68 and 69) we think will interest your students. You can develop many more. Give a copy to each of your students and then ask them to de-scribe what they see in the drawing or picture. Tell them also to indicate if they see a series of things.

Figure 68. Pictogram 1.

Figure 69. Pictogram 2.

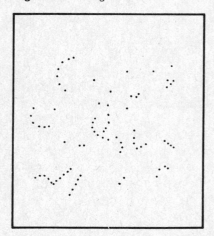

After the students have had sufficient time, you can engage them in several activities. At some point you will probably want to tell your students what the pictogram is. When you do, allow them some time to look it over again.

1. Discuss how they interpreted this diagram. Focus on patterns, differences, and likeness.
2. Use the diagram to interview other people (teachers, classmates, parents, brothers, and sisters).

Controlling Variables

A variable is something that may vary or change. In general, student experiments will be designed to test the influence of one variable on another. At first glance, this may sound simple. It is not! We suggest that you emphasize with your children that there are lots of factors (variables) that influence the behavior of a system. To illustrate this, try either the rocket or the melting ice activity with your children.

The Rocket

We recommend this be done outside! (Either that, or you'll have puddles throughout your room!) The rocket in this activity is an inexpensive plastic, water rocket available in most toy stores. Once outside put water into the rocket,

pump it a few times, and then fire it off. Have one of your students fetch the rocket, and then fire it off again. (Give it a few more pumps this time). Pose this question to your class: *What affects how high the rocket goes?* You'll probably generate a list something like this (bring large sheets of paper outside with you so that you can record the replies):

- the weight of the rocket
- amount of water
- the number of pumps
- how you held it
- how you release it
- the amount of wind
- size of the opening in bottom of rocket

If your students produce such a list, they have engaged in the process of identifying variables. Each factor listed above will have an effect on how high the rocket goes. The problem is finding how *each* affects the rocket's motion.

At this point, you should have rockets available for groups of three children. Ask your children which variable they would like to study to find how it affects the rocket's motion. Suppose they select the amount of water (which is the fuel). If you ask them how they will do this, you'll probably get responses such as the following:

1. Put in different amounts of water, give it ten pumps each time.
2. Lie on the ground, put different quantities of water in each time, but pump it the same number of times.

Notice in both responses that some variables are held constant, while the amount of water is varied. In this situation, water (fuel) is the controlled variable, while the height the rocket rises is the responding variable.

Now try it with your students!

Melting Ice

Give each of your students a plastic cup half full of crushed ice and a small baby food jar. Tell them they have five minutes to melt as much ice as they can.

Beware! This is noisy, chaotic, but fun. Students will subject the ice to a variety of melting techniques. They'll blow into the cup, wrap their hands around it, crush it up more, and so forth.

Melt that ice!

At the end of 5 minutes have the students stop and pour the water they melted from the ice into the jars. Find the champion ice melter!

Ask the children what factors affect the amount of ice melted. Here are sample responses:

· amount of ice started with
· size of the ice chunks
· size of the hands
· breath power
· temperature of hands
· time
· rate of crashing cup into table

Now ask which of the things listed above affects the amount melted the most. Any response the student makes is a hypothesis and should lead to an experiment. Regardless of the choice made, the student is going to have to keep all the other variables constant, and study the effect of that variable on the amount of melting. Repeating with other variables manipulated will enable him to accept, modify, or reject his hypothesis.

Generating activities that focus on the process of identifying and controlling variables is a wonderful way of getting children to begin to think experimentally. Table 13 may give you some ideas.

Table 13.

System	Inquiry	Variables Influencing the System
Potted plant	What will affect the rate of growth of the plant?	amount of water sunshine soil depth seed planted temperature of soil
Blotting paper	What factors influence how high water will rise on a strip of blotting paper?	width of paper thickness of paper texture of paper depth to which it is placed in water color of paper time held in water shape of paper
Paper airplane	Why does one paper airplane fly farther than another?	wingspan length of plane weight of plane distribution of weight initial thrust height thrown from
Stream table	What do you suppose causes the water to change speed?	amount of water flowing slope of stream obstacles in the stream shape of stream depth of stream
Storm (tornado)	What will affect whether we are hit by the tornado or not?	temperature of air location of fronts humidity wind motion pressure changes luck!

Hypothesizing and Designing Experiments

List A
· The earth is not flat.
· 1 inch equals 2.54 centimeters.
· Water boils at 100°C (in most locations).
· The density of this rock is 2.8 g/cc.

List B
· Girls are more intelligent than boys.
· Most metals are good conductors.
· Heavier objects fall faster than light objects.
· Rubber balls bounce higher than plastic balls.

Are these two lists alike? Or does each list contain very different types of statements. List A might be considered by most people to be a list of facts, while B contains statements that we may not know to be true. These are hypotheses. They are statements about a class of objects or events (metals, people, objects, balls). We might view them as hunches, guesses, or, put another way, statements we are not sure about.

Sensitivity

Here is an experiment you can try. Consider this hypothesis:

"Boys are more sensitive than girls."

First, we aren't talking about the psychological concept but rather the physical one! We want to find out to what extent boys are more sensitive to touch than girls.

To test this hypothesis we need to consider several things.

1. What do we mean by sensitivity?
2. How will we measure it?
3. How will we design an experiment to find out if boys are more sensitive than girls?

Sensitivity as used here is a concept that relates to the sense of touch. One's sensitivity can be measured by using a sensometer (Figure 70) to determine how close two toothpicks can be placed together before we cannot feel two points. The sensometer can be used to measure a person's sensitivity by touching the sharp ends of the toothpicks to various parts of our body (neck, forearm, palm, soles of feet, forehead).

Figure 70. How sensitive are you?

Designing the Experiment

Designing an experiment will involve:

· identifying and controlling variables
· selecting the test sample
· recording the data
· analyzing the data and drawing conclusions

There are problems you and your class of children can deal with as a group. Here are some possibilities to consider while doing this experiment.

Variables
- reading the sensometer accurately
- how lightly we touch with the sensometer
- sharpness of toothpicks
- age of students
- part of body tested
- tested sitting or standing

How to Get the Sample
- test all the boys and girls
- test the boys and girls in the fifth grade
- draw names out of a hat from boys and girls in entire school

Doing the Experiment

Ten boys and ten girls were selected randomly from both fifth grade classrooms (names were put in two hats from which they were drawn). Each person was tested for sensitivity on the right forearm and the sole of the left foot. The same person did all the testing. You can record the information on sheets like those shown in Figure 71. Have fun doing the experiment with your students.

Figure 71.   Sensitivity test record sheets.

BOYS
TOUCH TEST: forearm
distance between points

SAMPLE	TRIAL 1	TRIAL 2	AVERAGE
1			
2			
3			
4			
5			
6			
7			
8			
9			
10			

GIRLS
TOUCH TEST: forearm
distance between points

SAMPLE	TRIAL 1	TRIAL 2	AVERAGE
1			
2			
3			
4			
5			
6			
7			
8			
9			
10			

Designing an experiment involves very careful planning and consideration; however, it is not complicated or difficult. Boys and girls, particularly if you are teaching in the fifth grade or above, are quite capable of testing hypotheses by designing experiments. Look back at Table 13. Each of the inquiries posed in that chart requires that the student design an experiment. We hope you will spend time in group activities helping your children learn this fascinating skill so that they can answer their own questions.

Learning Centers

Children get turned on to learning if topics and ideas are highlighted, if they can choose from a number of alternatives, and if they can decide with whom they will work. Using learning centers in the classroom has certainly helped us meet these needs. There are probably as many ways to establish learning centers as there are classrooms and teachers. We hope you will use the Kid Stuff activities to develop your own learning centers.

Our intent in this section is to show how the processes of discovering, valuing, and exploring can be used to develop learning centers in your classroom. Here are two possibilities:

1. Combining the three processes around a concept or catchy phrase to form an integrated center.
2. Using the processes separately to establish discovering, valuing, and exploring centers.

Integrated Learning Centers

We have highlighted three processes we feel are important in helping children learn about science. We might consider them together in the scheme shown in Figure 72. This now gives us a vehicle to plan and organize our classroom around a particular concept, idea, or content area. Activities focusing on the processes of discovering, valuing, and exploring can be developed (and don't forget the 81 in this book) and integrated based on a content theme.

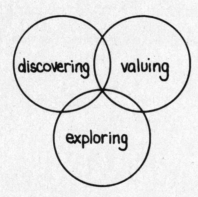

Figure 72. Integrating the processes.

Imagine developing integrated centers around the following themes:

DIG THE EARTH
MAPS, TRAVEL, AND ME
GO METRIC
BUGS AND OTHER LITTLE CREATURES
HOW HEALTHY ARE YOU?
MYSTERY LIQUIDS
VIOLENT STORMS AND WEATHER
HOW MUCH ENERGY?
LIFE IN THE POND
WHAT'S ON THE PLANETS?
STRANGERS FROM OUTER SPACE

Figure 73 is an example of an integrated center for Dig the Earth. We hope it will serve as a model for integrating a variety of other things.

Figure 73. An integrated learning center for Dig the Earth.

DIG

TEACHER STUFF

DISCOVERING

1. Observe different earth materials: rocks, minerals, soil.
2. Solve the inquiry problem: "Floating Rocks."
3. Use the sense boxes.
4. Make a scrapbook of prehistoric animals.
5. Can you classify these rocks?

VALUING

1. Make a sculpture using soapstone.
2. Write a fantasy story about a trip to the center of the earth.
3. Make pottery from clay.
4. Write a SYNTU about something in geology.
5. Make up a cartoon about "Spoiling the Land."

EXPLORING

1. Take a Yellow Pages field trip to a rock shop, geology department, and so on.
2. Experiment with rocks.
3. Make a mini-map of school grounds.
4. Collect earth materials and make a display.
5. Read about famous miners and geologists.

THE EARTH

KID STUFF

DISCOVERING

1. Use a hand lens and observe the materials in the box. Make up a game using the materials and play it with someone.

2. "Do rocks float?" Test these 3 rocks. Why does one float? (use pumice, sandstone, and granite).

3. What can you find out about the materials in these sense boxes? (In box 1, dirt for smelling; box 2, sand & gravel in bags for feeling; box 3, rocks to listen to but not look.)

4. Using the magazines make a science book of prehistoric animals.

5. Using the rocks and minerals box, identify these materials.

VALUING

1. Make a piece of art using soapstone and tools in the box.

2. Have someone read the fantasy story to you and then write about it.

3. Make something for yourself or a friend using the clay.

4. Write a SYNTU about one experience in dig the earth.

5. Make a cartoon about spoiling the land. Make a cartoon book with your friends.

EXPLORING

1. See the Yellow Pages and select an earthy field trip place. Invite them or go there.

2. Experiment with these materials to make some rocks: plaster of Paris, water, sand, gravel, shells.

3. Go outside and map the school area.

4. One weekend collect earth materials and make a map collection.

5. Find a book below or in the library.

The purpose of this center is to involve the students in many geology-related facets of the earth. These range from identifying rocks and minerals to making sculpture with soapstone. An awareness of some crafts, concepts, processes, and values related to geology will be major outcomes. The activities for each process can be arranged together in an accessible place in the classroom.

Process Learning Centers

The Kid Stuff chapters of Discovering, Valuing, and Exploring provide a second alternative for organizing and developing learning centers. The possibilities for activities and approaches are endless. We feel that process centers can be used to do at least three things:

1. arouse curiosity (discovery);
2. provide opportunities for the expression of feelings and emotions about science (valuing); and
3. unleash curiosity by allowing children to pursue science topics in depth (exploring).

Each process center could occupy a particular portion or corner of your classroom. You can heighten interest and motivation by changing the activities within each process center frequently.

The Discovery Center

This is the place in your room where the objects, materials, and things of the child's environment should dominate. The selection of activities should emphasize, as you recall from Chapter 7, sensory awareness, science processes, and inquiry problems. Activities for the discovering center could include any of the following possibilities:

1. Put on a blindfold and then reach into the touch box (feely box) to discover how each of the objects feels.
2. Mystery powders! What do you think these powders are? How can you find out?
3. Observing the aquarium. Each day for the next week look into the aquarium and look for changes.

You should also consider the activities in Chapter 7 for inclusion in the discovering center.

The Valuing Center

We have said over and over that this aspect of learning has been missing in science teaching. One way of making a commitment to humanizing your classroom is by reserving one part of your room for emotional, feeling, intuitive, and play activities.

Like the discovering center, you could coordinate topics in the valuing center with the topic your children are working on. The valuing center can be used to involve children in activities such as the following:

1. Graffiti board. Use a large sheet of oak tag and invite your students to write graffiti. Do you want to be very brave? Use the entire length of one of your classroom walls!
2. News reaction. Assign children to be responsible for bringing in science-related newspaper articles, headlines, cartoons, and editorials. Post them in the valuing center and then solicit your children's opinions on reaction sheets placed next to the news.
3. Fantasy tapes. Make an audio tape designed to stimulate your children to be creative and imaginative. Provide several segments on the tape, each of which could be considered as separate and alternative activities.

Some examples would be:

· Record a popular song; at the end ask the children to draw, paint, or write what the song means to them.

· Record sounds of nature (birds, wind, storms, people, insects) and ask the children to write a story, play, or poem about them.

The activities in Chapter 8 provide an additional and rich source of ideas for your valuing center.

The Exploring Center

Imagine a section of a classroom devoted to the construction of projects, suggestions for local field trips, ideas for interesting experiments to do, and a wealth of reading, listening, and visual material. The exploring center in your classroom should be designed to do these things.

Although exploring is the last of the three major processes, we've considered it as also a beginning, a beginning of a search that never ends; a search for ways of explaining our world and enjoying it; a search that combines our full mental powers and abilities to think logically with the affective and aesthetic side of our humanness. We explore so that we can discover and value once again.

Conversation

Joe: Well Jack. I guess that's just about it.

Jack: I'm glad we end with exploring. Exploring really is an optimistic process isn't it?

Joe: Right. When we explore we press onward in search of newer things to do and to learn.

Jack: I wonder what's out there.

Joe: What do you mean?

Jack: What's out there? What will the kids explore? I mean beyond the classroom.

Joe: Well that's the question that no teacher ever finds the answer for. All we can do is help children explore while they are under our care. We never know where that ability for exploration will lead a child as he becomes an adult and goes through life.

Jack: That's one of the frustrations of teaching— never knowing what direction our "children" eventually do take.

Joe: All we can do is try to help them become all they are capable of becoming.

Jack: If we do the hard work of teaching and provide a learning environment for science and for all areas of the curriculum that may help them see themselves as people in progress, maybe they'll be okay.

Joe: They'll be okay.

Notes

CHAPTER ONE

1. Bruno Bettelheim, *Love Is Not Enough: The Treatment of Emotionally Disturbed Children* (Glencoe, Ill.: Free Press, 1950).
2. Robert L. Shannon, *Where The Truth Comes Out: Humanistic Education,* © 1971, Charles E. Merrill Publishing Company, Columbus, Ohio, pp. 14–15. Used with permission.
3. Robert L. Shannon, *Where The Truth Comes Out: Humanistic Education,* p. 27. Used with permission.
4. Reprinted with permission of Macmillan Publishing Co., Inc., and Turnstone Press Ltd. from *Jonathan Livingston Seagull* by Richard Bach.
5. Reprinted with permission of Macmillan Publishing Co., Inc., and Turnstone Press Ltd. from *Jonathan Livingston Seagull* by Richard Bach.
6. Reprinted with permission of Macmillan Publishing Co., Inc., and Turnstone Press Ltd. from *Jonathan Livingston Seagull* by Richard Bach.
7. Reprinted with permission of Macmillan Publishing Co., Inc., and Turnstone Press Ltd. from *Jonathan Livingston Seagull* by Richard Bach.
8. Earth Science Teacher Preparation Project, from the "I-Level Game" in the *Gift Garden of FantaSeeds,* © copyrighted and packaged for the American Geological Institute, 1974. Funded by the National Science Foundation. The FantaSeeds contain some of the best ideas around for helping us humanize our classrooms.

CHAPTER TWO

1. Thomas A. Harris, *I'm OK—You're OK,* copyright © 1969 by Thomas A. Harris, p. 221. With permission of Harper & Row, Publishers, Inc.
2. From the song "Everything That Touches You," written by Michael Kamen, as recorded by Bonnie Raitt on Warner Brothers Records. © Mother Fortune Music, Inc. Used with permission.
3. This game is based on a chapter in David Weitzman's book *Eggs and Peanut Butter,* copyright 1975, Word Wheel Books, Inc., Menlo Park, California. Used with permission.

CHAPTER FOUR

1. Wall plaque at the entrance of the John F. Kennedy School, Winooski, Vermont.

CHAPTER FIVE

1. Carl R. Rogers, *Freedom To Learn,* © 1969, Charles E. Merrill Publishing Company, Columbus, Ohio, p. 228. Used with permission.
2. In the teacher materials for *Essence I,* Addison-Wesley Publishing Company, © 1971, the American Geological Institute, p. 29. Used with permission.
3. Based on a questionnaire developed by the Earth Science Teacher Preparation Project, sponsored by the American Geological Institute and funded by the National Science Foundation.
4. The items listed in this instrument are only suggestions. Chapter 8 is a rich source of additional questions, particularly the activities Science Soapbox and Project Work Using Sticky Science Issues. We suggest you form your own instrument using some or all of these items in addition to your own.
5. Reprinted with permission from the SCIS Evaluation Supplements, written and published by the Science Curriculum Improvement Study. Copyright © 1972 by the Regents of the University of California.
6. From *Intermediate Science Curriculum Study, Individualized Teacher Preparation, Questioning,* pp. 3–5 and 3–6, Silver Burdett Company, Morristown, N.J. © 1972, The Florida State University. Reprinted by permission of The Florida State University.
7. Patricia Blosser, *Handbook of Effective Questioning Techniques,* © 1973, Education Associates, Worthington, Ohio, p. 86. Used and modified with permission.
8. From *Intermediate Science Curriculum Study, Individualized Teacher Preparation, Questioning,* p. 2–2, Silver Burdett Company, Morristown, N.J. © 1972, The Florida State University. Reprinted by permission of The Florida State University.
9. Joseph Abruscato, The Student Communication Analysis (SCAN) System: Its Development and Preliminary Utilization, doctoral thesis, Ohio State University, Columbus, Ohio, 1969.
10. Earth Science Teacher Preparation Project, from the "I-Level Game" in the *Gift Garden of Fanta-Seeds,* © copyrighted and packaged for The American Geological Institute, 1974. Funded by the National Science Foundation.

CHAPTER SIX

1. Maxine Greene, *Teacher As Stranger,* © 1973 Wadsworth Publishing Company, Inc., Belmont, California, p. 282.
2. Ashley Morgan, "An Invitation to Learning to Learn and Grow," in *Science Education for Man: Learning to Learn, Learning to Grow* (Atlanta, Ga.: Center for Improving Elementary School Science, Georgia State University, 1974), p. 77.
3. Ashley Morgan, "An Invitation to Learning to Learn and Grow," p. 77.
4. David Hawkins, "Messing About In Science," *Science and Children* (February 1965): 5–9.
5. Robert Samples, "Toward the Synergic School Room," Essentiasheet No. 3 (Olympia, Wash.: Evergreen State College, Fall 1974).
6. Reprinted from *The Prophet,* by Kahlil Gibran, with permission of the publisher, Alfred A. Knopf, Inc. Copyright 1923 by Kahlil Gibran; renewal copyright 1951 by Administrators C.T.A. of Kahlil Gibran Estate, and Mary G. Gibran.

CHAPTER SEVEN

1. J. Bronowski, "Science in the New Humanism," *The Science Teacher* (May 1968): 13.
2. Gloria Castillo, *Left Handed Teaching,* © 1974 Praeger Publishers, New York, p. 153
3. Gloria Castillo, *Left Handed Teaching,* p. 153.
4. Gloria Castillo, *Left Handed Teaching,* p. 86. Used with permission.
5. The games are based on two Elementary Science Study Units, Rocks and Charts, © 1967 Education Development Center, Inc., and Attribute Games and Problems, © 1967 Education Development Center, Webster Division McGraw-Hill Book Company. The publication is not endorsed by the original copyright holder.
6. *Investigating the Earth, Teacher's Guide* Part 2. Houghton Mifflin Company, Boston, © 1967 American Geological Institute, p. 784. The publication is not endorsed by the copyright holder.
7. Based on the Elementary Science Study Unit *Clay Boats,* Webster Division McGraw-Hill Book Company, © 1969, Education Development Center, Inc. All Rights Reserved. The publication is not endorsed by the original copyright holder.
8. This technique for stimulating children's inquiry is based on the work of J. Richard Suchman and more fully developed in the Inquiry Development Program, © 1966 Science Research Associates, Inc., Chicago, Illinois.
9. From *Inquiry Development Program,* Idea Book by J. Richard Suchman, © 1966, Science Research Associates, Inc. Reprinted by permission of the publisher.

CHAPTER EIGHT

1. Both poems were written by students in Margaret Carraway's creative writing enrichment class.
2. Harold C. Lyon, Jr., *Learning to Feel—Feeling to Learn* (Columbus, Ohio: Charles E. Merrill Publishing Company, 1971), p. 18.
3. This is one of the best activities for introducing your kids to the environment. It is one that Ted Colton provided us. The Creepy Crawly Race is also his idea.
4. George I. Brown, *Human Teaching For Human Learning* (New York: Viking Press, 1970) p. 46.
5. Modified and based on *Essence I,* Addison-Wesley Publishing Co., Menlo Park, California, © 1971 American Geological Institute. Used with permission.
6. Reprinted from *Latest Will, New and Selected Poems,* by Lenore Marshall. By permission of W. W. Norton & Company, Inc., and McIntosh and Otis, Inc. Copyright © 1969 by Lenore Marshall.
7. Joan Goodwin, *Human Heritage,* Part One. Copyright © 1971 by Beacon Press. Published by the Unitarian Universalist Association. Used with permission of the publisher.
8. Earth Science Teacher Preparation Project (ESTPP), from "Valuing" in the *Gift Garden of FantaSeeds,* © copyrighted and packaged for the American Geological Institute, 1974. Funded by the National Science Foundation. Used with permission.
9. Louis Raths, Merrill Harmin, and Sidney Simon, *Values and Teaching* (Columbus, Ohio: Charles E. Merrill Publishing Company, 1966).
10. These were developed by students in Bob Jaber's science class. We thank them all.
11. Adapted from "Action Learning" by Sharon Lehman, in the *Gift Garden of FantaSeeds,* Earth Science Teacher Preparation Project (ESTPP), copyright © 1974, The American Geological Institute. Used with permission.

CHAPTER NINE

1. The idea for mini field trips is based on the 10 minute field trip idea developed by the Environmental Studies Project; now Essentia, Evergreen State College, Olympia, Washington.
2. H. Poincaré, *The Values of Science,* translated by G. B. Halsted (New York, N.Y.: Science Press, 1907), p. 8.

Photographic Credits

The photographs in the book were taken by Jack Hassard, except for photographs on pages 36 (lower left), 79 (upper left), and 103 by Ted Colton, on page 4 by Bob Jaber and on page 87 by Gary Meador.

The artists: The pictures at the beginning of each chapter in this book were created by these talented boys and girls in Jane Hannon's and Carolyn Hinsucker's classes.

Notes About the Authors

Joe Abruscato and Jack Hassard are an effective and active science education team. They are former teachers who are now assisting preservice teachers, inservice teachers, curriculum coordinators, and administrators in many states who are interested in improving science education experiences for children and youth. Their activities have included developing and leading workshops for teachers and other school personnel, teaching preservice and inservice courses at the undergraduate and graduate level, speaking at conferences, and writing articles for periodicals.

Through such experiences they not only have shared their own skills and knowledge but have also received many ideas and activities from classroom teachers. *Loving and Beyond: Science Teaching for the Humanistic Classroom* is a natural extension of this sharing process. It has been specifically designed to tie together important ideas about how to organize humanistic science experiences with an extensive collection of practical activities that can easily be used in the classroom.

Thank you for using our book. We have worked hard to put it together for you, and we would really like to hear from you. If you have any reactions (good or bad) that could help us, please let us know. Also, if you have ideas about teaching children, or great science activities that you would like to share with others, we will try our best to include them in future projects.

Thanks again!